NATURE,
THE TRUE GOD

Updated Evidence for Unity

PON SATCHI

BALBOA.PRESS
A DIVISION OF HAY HOUSE

Balboa Press books may be ordered through booksellers or by contacting:

Balboa Press
A Division of Hay House
1663 Liberty Drive
Bloomington, IN 47403
www.balboapress.com.au
1 (877) 407-4847

Because of the dynamic nature of the Internet, any web addresses or
links contained in this book may have changed since publication and
may no longer be valid. The views expressed in this work are solely those
of the author and do not necessarily reflect the views of the publisher,
and the publisher hereby disclaims any responsibility for them.

The author of this book does not dispense medical advice or prescribe the use
of any technique as a form of treatment for physical, emotional, or medical
problems without the advice of a physician, either directly or indirectly. The
intent of the author is only to offer information of a general nature to help
you in your quest for emotional and spiritual well-being. In the event you use
any of the information in this book for yourself, which is your constitutional
right, the author and the publisher assume no responsibility for your actions.

Any people depicted in stock imagery provided by Getty Images are
models, and such images are being used for illustrative purposes only.
Certain stock imagery © Getty Images.

Print information available on the last page.

ISBN: 978-1-5043-2051-1 (sc)
ISBN: 978-1-5043-2052-8 (e)

Balboa Press rev. date: 01/25/2020

Dedicated to my parents.

Contents

Preface

This book is written to share the recently discovered facts that science and history have uncovered in the last few years regarding the events in the creation of the universe. I hope this is made easy for the freethinkers, our younger generation, and those who genuinely seek the truth. Understanding the truth about creation and the actual Creator will simplify our lives and its quality when it comes to peace, happiness, problem-solving, time-saving, and maintaining unity.

Nationalism, tribalism, and religious fanaticism are frightfully destroying our unity, security, and peace. When you don't belong to any of these identities, you become simple, free, and harmless. There is only one truth, while there are many—some think too many—contradictory beliefs. Understanding and following the path of truth results in a single belief and a single God. In fact, we believe something that we do not know for certain or have been asked by someone to do so.

Reading this is book in full provides the right answers to most of the conventional questions that we come across in life, such as, Who am I? Who is God? Why are we born? What is true life? How do I live a complete life?

I pre hose who genuinely want to be informed about the latest discoveries in science and history. It is not intended to hurt, convince, or convert anyone to these findings. I would rather like the freethinkers to just read with open and inquisitive minds and genuinely investigate the truth based on the given evidence. In fact, complete changes to beliefs and traditions may take several years or generations, but merging slowly into the truth can be easy, allowing one to reap the benefits eventually. Always better late than never.

I share the latest and most suitable big bang theory of creation and Darwin's evolution theory. These two are independently formulated and have been established all over the world. Several scientists, with the help of modern supercomputers and powerful telescopes, have confirmed the principles of these theories.

I have also tried my best to collect and incorporate the required basic evidence from all the fields of study, such as physics, chemistry, cosmology, history, biology, palaeontology, geography, geology, genetics, medicine, and theology. In order to simplify, I prepared tables, flow charts, images, and timelines so that everyone can easily understand the chain of events that have happened and are happening in the universe and on our planet.

Acknowledgements

Thanks to all the scientists and experts who published research papers, images, and lectures via websites. Thanks to my wife Jega and my daughter, Kalpana, for initial editing work and support in writing this book.

Introduction

For most of us, the beliefs in our heads are based on where we were born, how we were raised, and who influenced us. Exceptions are the very few fearless freethinkers who are open and not prejudiced, and live a realistic life. There are several beliefs and differences in opinion all over the world; the truth is just one of them. We are free to choose any, but we should remember that we are not free from the consequences of our choices!

In ancient days, people believed that the earth was flat. But in 600 BC, Pythagoras first stated that the earth is spherical, and by 340 BC, Aristotle confirmed it. Other scientists later confirmed this truth by observing eclipses and found the circular shadow of the earth fell on the moon during these eclipses. During ancient days, people also believed that the earth was in the centre of the universe, and the other planets and the sun rotated around the earth (geocentric model). But in 220 BC, Greek astronomer Aristarchus first stated that the sun is in the centre, and the planets, including our earth, were going around the sun (heliocentric model). People slowly accepted these facts and changed their beliefs, even though it went against their original beliefs.

At birth, Jesus was not a Christian, Buddha was not a Buddhist, and Muhammad was not a Muslim. They all brought their new faiths to the people modifying their past beliefs. So why do we not also update now, when, after a long time, additional facts are being discovered. True wisdom will accept the truth and reality in order to be smart and advanced.

When kings ruled during ancient times, there were too many humanitarian problems due to egoistic wars between the kingdoms and tribes. There were no international accords like there are now. Religious priests played dominant roles in advising the kings. The minority tribes with other beliefs suffered in slavery and harsh treatments. As such, the prophets and messengers of God came from the affected groups and volunteered to help their people. In order to convince and unite the people, these religious authorities used several unknown and unseen concepts, like "God," "angels," "revelations," "sin," "hell," "heaven," "paradise," "soul," "spirit," and "fate." This invoked fear and disciplined the people to keep them under control—to a certain extent. Those who were very close to the prophets became their disciples and continued the spread of their faiths by writing holy books and sermons. Some others went around in large groups as an army and forcibly converted people by using threats and violence. Those who resisted were killed. This is the brief history how religions originated and spread.

During ancient time, establishing new religions was sort of a makeshift answer to an urgent need. The prophets came with their visions and dreams to claim that God had spoken to them and revealed facts in their dreams or in person. The people had to believe and follow without questioning.

Unfortunately, facts about our universe were not available to them because they did not have modern tools and facilities to explore. So they had to assume concepts or make up imaginary stories to convince people. If analysed well, every concept of theirs was based on giving importance to only the human species. Images of God were also projected as having human qualities based on human needs and desires. As such, a spontaneous creation theory was assumed starting with human species and others to support. However, we cannot blame the good attitudes of our prophets who genuinely wanted to help the people.

But now, when so much of information is available everywhere, an educated and civilized person should not stagnate with any baseless, ancient, false beliefs. We should not act simply on the claims of the ancient scriptures that were meant for our ancestors. We are supposed to proclaim no faith on impossible events and secret revelations. We also have to be rational and look for evidence and be satisfied with our own consciences.

When there is an urgent need to update and upgrade, it is sad to see that some religions are just claiming their superiority and dominance.

At the moment, we cannot ignore our responsibilities to take a step out to guide our next generation in the right path. We have too many intelligent and smart children who have started to question our assumed ancient beliefs and superstitions. Information technology, at present, has taken them deeper into searching for facts. Unfortunately, we are unable to genuinely answer their questions satisfactorily.

In order to update facts about creation and the Creator, I like to share the concepts of the big bang theory of creation and the natural selection theory of evolution by Darwin. I found both theories, one about the nonliving (matter) and the other about the living (life), complement each other regarding science and history. Both theories merge well to describe the development of matter and life in the universe and on our planet.

It is remarkable to observe that, even though these two are independent theories from different fields of science and from different time periods, they support each other because they are true. Both merge very well in the timeline of history regarding changes in the environment of the universe and earth. Both also follow the same efficient but simple mechanism of cause and effect or the domino effect. In addition, both theories answer with reason any related questions about the past, present, and future of our universe and earth.

Unlike revelations, the big bang theory is a product of the hard work of thousands of scientific brains that researched for hundreds of years. This theory was challenged, confirmed, and put together after using modern supercomputers and powerful telescopes. This theory is well established and accepted all over the world. If you accept science, you can easily trust the big bang theory!

Chapter 1
General

Why Should We Review and Update Ancient Concepts?

No one can deny that in the last fifty to sixty years, the marvels of modern science have changed our lives very fast in the most unimaginable ways. Humans are now nature's detectives and are supposed to inquire about the developments in the cosmos and on our planet. Life has become accessible and quite easier now than it was sixty years ago. But still, we have not fully updated many important and essential facts about our universe. Updating facts will enhance natural perception and consciousness, which is very important for better survival and peace in life.

It is not right to live with falsehoods or conceal truths when you have the opportunity to know the reality. Our belief in the concept of God and creation is very ancient and has no proper evidences. Modern computers and powerful telescopes have uncovered new findings about our universe and the planet where we live. However, in many other fields—like food, dress, transport, communication, and habits—we have regularly updated and upgraded information to add comfort to our lifestyle.

Another important reason to review is for our peace and safety, which are going down the drain badly due to varying beliefs. This is leading us to misunderstandings, egotism, greed, and power seeking. As such, the world is in major turmoil and now

1

filled with acts of terrorism, wars, drugs, refugee exodus, and so on. We have lost our freedom to move about, lost trust in fellow humans, and lost love and unity. This is because we are holding on to ancient beliefs that were meant for the past. That is the main cause of our division and increasing enmity.

But other factors also divide us. They include race, colour, culture, and language. But the most dangerous is religious fanaticism. People are trying to please their Gods and want to get to the unknown and unseen heaven by taking a shortcut. We also waste a lot of our lifetimes, energy, and wealth with these unknowns. We build too many extravagant houses of worship and weapons of eradication instead of helping the needy.

At this rate, the human race has only a very short time to exist on our planet. We should not forget the past global extinctions, especially of the dinosaurs during the prehuman era about 65 million years ago. But our problem is more serious and dangerous because we are trying to eradicate each other using our own discoveries, like nuclear power and other weapons. Increase in drug usage is another serious problem facing humanity these days when the truth is ignored.

Our brains have developed, but our hearts have shrunk. When there is only one truth, there is no reason to have so many faiths and divisions among the faiths, and to fight each other. Next we review all our beliefs for the truth and develop one concept. It is high time to learn the facts and genuinely accept the single truth, shedding our egos and become peaceful for a united society. If not, it is certain that our future generations will also suffer very badly and may go extinct.

A Brief Review of the Origin of Some Faiths

The Hindu religion is said to have started about five thousand years ago or more as a kind of civilization in the Indus Valley and along the banks of the river Sind. Hinduism is more of a philosophy and a way of life than a religion.

The pioneers of Hinduism are the great sages, yogis, and mystics. Hinduism divides the history of the universe into four yugas, totaling about 4.32 million of years. Followers of Hinduism believe that God incarnates as powerful men from time to time to eradicate the bad people (demons). They also believe in rebirth, soul, sin, fate, heaven, and hell. As time passed, they imagined a family set up for God, like humans have, and added several additional related member deities for various functions.

Key principles of the religion are based on dharma (righteousness) and karma (actions). The spiritual books in Hinduism range from the complex writings of the Vedas (knowledge about divine) and Upanishad (fundamental teachings on concepts like karma, samsara, moksha, atman, Brahman, yoga, and meditation) to religious epic stories for people help people understand the belief. Hinduism emphasizes daily practices of yoga for the body and meditation for the mind.

There are about three divisions in Hinduism, namely Vaishnavites (70 per cent), Shaivites (27 per cent), and Neo or reformed Hindus (2.6 per cent). In total, there are about 1.1 billion Hindus in the world, most of them living in India.

Judaism came about four thousand years ago, when God asked Abraham and Sara to leave Mesopotamia and migrate to the

3

land of Canaan. Further, it was claimed that God promised Abraham, via revelations, that with the land of Israel as his tribe, they would be the chosen people.

But when the Egyptians invaded Israel, the chosen people were enslaved. Later, Moses freed them with the help of God performing miracles. Moses also received the tablet containing the Ten Commandments from God and gave it to his people, who were in moral dilemma.

There are about 14 million of Jews in the world (0.23 per cent). About 6 million currently live in Israel. There are five divisions now in Judaism (Conservative, Secular, Reform, Orthodox, and Reconstructionists). The Tanakh (often referred to by non-Jewish people as the Hebrew Bible) and the Torah are their holy books.

Christianity came from Judaism two thousand years after Jesus. The Romans terrorized the Jews and crucified Jesus for the divine sermons and the miracles he performed. The key virtues in Christianity are love, sacrifice, and forgiveness. The disciples of Jesus continued the spread of Christianity all over the world with the Bible.

In 1517, Martin Luther, a German monk, reformed this religion by splitting Western Christianity into two: Roman Catholic and Protestant. Then by 1534, Henry VIII, the king of England, initiated this reformation in his country. Now there are more than eight divisions within Christianity (Catholics, Orthodox, Conservative and Liberal Protestants, Pentecostal, Anglican, Jehovah's Witnesses, African indigenous). There are about 2.38 billion Christians (31.2 per cent); it is the largest religion in the world.

Buddhism was established 2,500 years ago based on the philosophies of Lord Buddha, who emphasized meditation, surrender, and good conduct to attain Nirvana (enlightenment). Buddha condemned all superstitions and the false beliefs of his parents' religion. He sacrificed his kingdom and walked away in search of spiritual teachers to find peace. Buddhism now has many subdivisions, but the major three branches are Mahayana (East Asia—Japan, China), Theravada (southern Asia—Sri Lanka, Burma, and Thailand, and Lamaism or Vajrayana (Tibet and Nepal). Buddhists have no belief in a personal God. Buddhists make up about 6.9 per cent of the people in the world.

Islam came in via the prophet Muhammad about fourteen hundred years ago. It was established when people in Arabia were experiencing serious problems, suffering and dying from a never-ending Byzantine-Sassanid War that produced famine and a fatal plague in the region. The prophet Muhammad was involved in three battles between the Muslims and the Makens from AD 625 to AD 627. Muslims won the first and the third, followed by the treaty in AD 628. The march on Makah happened about AD 630.

Islam means peace or submission to the will of God. Devotion to God and the Brotherhood of Man are the key teachings. The Holy Quran is their Holy Book. Today, there are four divisions of Islam: Sunni 84 per cent, Shiite 15 per cent, Ahmadiyya, and Druze. There are about 1.8 billion Muslims in the world.

About Beliefs, Philosophies, Logic, Intuition, and Science on Truth

Beliefs are complex and determined. True knowledge is seldom found in beliefs. You believe something when you are not 100 per cent sure about it, but someone else says it is true. Our parents are the first people to pass on their beliefs to us. Most of us keep these, especially the religious beliefs, throughout our lives and pass them to our future generations as a tradition or culture. Rarely are these religious beliefs reviewed because we fear God. Later, teachers and religious authorities can influence our beliefs to change.

Philosophical explanations can be interesting and very thought-provoking. They imply many things but cannot always prove or disprove difficult theories. Most of the time, philosophers give fantastic explanations about the things that you do not know and have not experienced.

Logic is only useful to establish something that you already believe to be true or untrue. Logic is also a means to support the assumptions and presuppositions we have. If we are skilful, we can use logic to convince other people. But logic seems to be convincing only at face value. An equally skilful person can make a contrary point, winning over the truth. With logic alone, one can never establish the truth.

Rational thinking involves complementary functioning of the human mind. Rational thinking is linear, focused, and analytic. This treats all living things as a type of machine and the universe as one huge mechanical system, operating according to exact mathematical laws of space, time, and matter.

Intuitive knowledge, on the other hand, is based on a direct non-intellectual experience of reality. It is proclaimed to arise from an expanded state of awareness and enlightenment. So intuitive knowledge is always synthesising, holistic, and nonlinear. Others cannot measure the level of intuition in a person to be fully convinced of something's veracity. For most people, it is not an acceptable way to understand or explain.

Unfortunately, there are too many limitations on the above methods of reasoning or discussion. Based on these, nothing will help us to find the truth, especially on the topic of creation and the Creator. As such, we have to seek real ways to learn the truth: scientific and historical.

Science Is True Whether You Believe It or Not

Scientific explanations are presented after several enquiries and observations. They are confirmed with repeatable experiments and evidence. They do not have any standpoints for or against a belief.

Science does not depend on beliefs. Instead, it investigates and tests theories or hypotheses to confirm whether they are true. Science is rational and not vested in one outcome. Science is open to alternatives and is a discipline that is prepared to examine and upgrade its findings and conclusions.

Science is a pathway to finding the truth. Scientists discover facts from natural events and related laws, and then test them with experiments for inferences and conclusions. Scientific laws are actually nature's laws that are discovered after scientific research.

A scientific theory is formulated from facts. It is more than a hypothesis because it is normally challenged and tested rigorously with repeatable experiments. Science properly considers all evidence and follows its findings to wherever they lead, without preconceived conclusions. Fundamentals of science do not change, but the finer details may change with new evidence and discoveries.

Science is never against any religion; nor does it support atheists. Instead, it follows evidence and valid logic, which is why people say that "Science is always true whether you believe it or not."

Classical science has driven Western society to believe that life is a competitive struggle for existence, and progress has to be achieved through technological advancement. Three laws of motion by Newton and the laws of thermodynamics (law of conservation of matter and energy and the law of increased entropy) were the important successful theories during this period. Classical science stood for absolute certainty.

Modern science came in after the discovery of radioactivity in 1986 by Henry Becquerel and cracked the world of classical science that dominated till then. The great scientist Albert Einstein ignored the limitations of classical science and presented the theory of relativity in 1915 that went beyond the three dimensions (length, width, height) to include time and space.

The theory of relativity also describes gravity, which governs the motion of the larger objects like stars and planets. Thereafter, the atom was discovered by Earnest Rutherford, who found that the atom is 99.9 per cent emptiness (space), and its nuclear system is somewhat similar to that of the solar system. He claimed that the electrons move around the protons in orbits, in a way defying Newton's laws of motion.

Werner Heisenberg later discovered the principle of uncertainty. The theory of quantum mechanics deals with the world of subatomic particles. Quantum physics has helped us to look closer into the stars and galaxies and to understand the structure and composition of the universe using quantum computers.

The theory of relativity, quantum mechanics, and the uncertainty principle form the foundation of modern science. This boosted our knowledge in transcending the limitations imposed by our five senses. Here we can also see that modern science stands for the impossibility of absolute certainty.

This book explores in some depth religion and science to answer the great questions we have as humans in trying to understand our creation and the Creator.

Chapter 2
Creation of Matter

Ancient Creation Theories

Several theories of creation have been put forward by different beliefs and religions over time. None of these are complete and cannot answer important questions. There are many assumptions, made-up stories, and even fantasies with no evidence to support them. The general premise of these theories is that of spontaneous creation with God as the Creator. These theories would have helped our ancestors at that time but are not relevant for today. Unless we are willing to update and upgrade our understanding, we cannot answer those many until-now unanswerable questions. Essentially, the two dominant theories about creation boil down to:

- Spontaneous creation theory by religions, where a God sitting in heaven, as the Creator, dropped fully developed living and nonliving (matter) to earth. Here it is assumed that God designed the universe and its members, intelligently including every finer requirement. Also, God is always involved with us and the changes happening every moment, keeping accounts of our actions—sins and good deeds.

- Scientific evolutionary creation theory, in which nature is the Creator. The universe is naturally developing and evolving with time on a cause-and-effect mechanism. In this case, the entire universe acts as the Creator (God), and everything in the universe is connected to each other in some way.

1. One of the oldest theories of creation is that the whole universe is nothing but an illusion, a dream, a projection, a hallucination, or a figment of the imagination. It has come from nonexistence to apparent existence, and we are experiencing a world of plurality. This delusion is something similar to the vision of mirages in the desert. This theory indirectly points to the unavailability of information and evidence in the ancient time to understand or explain creation. When matter is visible and distinct to our senses, calling this universe an illusion is far-fetched.

2. The most well-known creation theory believed by most people is found in Genesis. It says that God created the world in six days and rested on the seventh. First, he created the day and night and then the sky, sea, and earth. Then he created the sun, moon, stars, and animals. The last to be created was the first man, Adam, whom God created in his image, and then the first female, Eve, from Adam's rib. But for non-believers of this theory, there are many unanswerable questions regarding the calendaring of days (seven days a week by God), the order of creation, and about the millions of species. Also this story does not add up to the actual age—billions of years—of our planet.

3. Vedanta considers the world as a potential tree in a seed, which is previously not manifested because of Maya, beyond the senses, the mind, and the intellect. The delusive play of Maya clears when right conditions are available, and then creation happens. Vedanta insists that the plurality of the world is not real but a superimposition of the real. It is just like how in a semi-dark room, a long rope can appear as a snake because of misapprehension. Here the

rope did not create the snake, yet the snake cannot be seen without the rope. So Vedanta always wants us to remove the ignorance so that wisdom can identify the truth or reality. Though this theory aligns with the fact that the universe is constantly changing, it fails to explain creation fully and with certainty.

4. Another theory is that the world has come from a process of modification of the Infinite (God), who has become the whole universe. The doubt it has is about the deterioration of the quality of the original God after getting into bits and pieces. This looks all right in the outline, but falls short in explaining the mechanism of conversion.

5. Some other creationists believe that creation is just the will or play of God, and time is the cause for these manifestations. Here at least two important questions arise in relation to the desires of God, who is supposed to be absolute and free of desires and wants. The second being the reality of time, which is unacceptable when infinity has neither a beginning nor an end.

6. Some non-creationists believe that the world is a long dream, and everything appears real until the dream or illusion ends. They argue that an unreality cannot strike our perceptions. But if they can generate feelings, such objects have a reality, even though they are short-lived.

7. A major school of Buddhists, the nihilists, believe that this existence of the world has arisen from non-existence. This is very difficult to hold as something cannot come from nothing unless magic happens. Also, no mention was given to the creation of the universe.

Modern Creation Theory

The big bang theory is the most suitable and the latest prevailing cosmological model and explanation for the creation of the universe. This theory answers all question to the most satisfaction compared to other fantasy creation stories. The big bang theory is no longer hypothetical because it can be directly tested in a laboratory by recreating the conditions of the big bang by colliding atoms in the atom smashers.

The initial spark for this theory came in 1927 from a Belgian Catholic priest, Georges Lemaitre, who was also a scientist. He proposed that the universe was in expansion. This gradually led to the present big bang concept, which was confirmed in 1990, after the Hubble telescope was launched.

This big bang theory was derived after several scientific researches, laboratory experiments, and mathematical calculations using ultra-modern computers and the latest powerful telescopes. Using general relativity, scientists worked from the present backwards and forwards in the time channel.

Scientists and historians have explained the creation of nonliving matter via the big bang theory and the evolution of living organisms on our planet through Darwin's theory of natural selection.

The big bang theory states that, after a huge explosion about 13.7 billion years ago, this universe emerged from a tiny singularity (primeval body). According to modern science's big bang theory) and true history, the initial cosmic egg was there as an infinite one (God). The creation design occurred naturally, with scientific reasoning in the most intelligent way that did not require any external support or a director.

The blast brought out all the matter and the energy of the whole universe in the form of subatomic particles and energy in the form of heat. As time passed, expansion and cooling took place. The subatomic particles fused to form the simple elements hydrogen and helium.

Hydrogen, helium, and other debris clumped together through gravitational force in groups to form billions of stars. These stars clustered together, again by gravity, to form the galaxies. Our galaxy is called the Milky Way. Our sun is a second-generation star. Our solar system is located in the outer region of our galaxy, where there are minimal radiation and other electromagnetic interferences. The next sections explore the big bang theory in more detail.

The Big Bang Theory of Creation: Overview

As mentioned previously, the big bang theory states that the universe came from a tiny "cosmic egg," or a primeval body (singularity) that exploded. The explosion brought in the matter and the energy of the whole universe. Soon after the explosion, very high temperatures and a soup of subatomic particles (quarks) were produced that needed more space, and expansion started immediately. Time also started from this point of explosion.

Discoveries in astronomy and physics have shaped the history of the universe and shown beyond a reasonable doubt that this universe began about 13.7 billion years ago and an end many billions of years after.

The billions of years that the universe has existed is generally mistaken for eternity. But the big bang theory scientifically and historically explains clearly the creation of our universe occurred after that explosion. The expansion and gradual cooling caused the formation of the various cosmic objects that we know today. Astronomers, through a phenomenon known as cosmic microwave background radiation, are still witnessing the "echo" of the bang and expansion.

Time and space started together from the moment of the bang. The time is specific to the present universe.

To study and confirm the big bang theory, similar but smaller-scale experiments are being performed by scientists from Europe, the United States, Japan, and Canada in the large laboratory CERN in Geneva using the Large Hadron Collider.

From events after the big bang, it was found that this primeval body contained all the energy and the matter for the universe in a very condensed form. The cause for the big bang was the severely compressed energy and the matter in a tiny body that could not hold them any longer.

Many people are unable to understand the origin of this initial tiny body, the singularity! In fact, this came as the end product of the previous cycle of the universe. The answer to this question will be better understood if one reads through the whole events of the cycle of the universe up to the end—after the Big Crunch—to learn how it develops.

Following the big bang, the changes and new events occurred, and are occurring in the universe every moment, like a domino effect. This is an automatic process, happening in order according to the laws of science based on a cause-and-effect mechanism. This is supposed to be the most intelligent and efficient system to go on. The recycling of the matter and the energy in the universe is an added intelligence to be independent and makes the universe self-sufficient.

The process of fission and/or fusion does the recycling. During the end of the cycle of the universe, the used energy and debris of the matter are sucked in by the black holes, squeezed, and made tiny and dense by the energy from the dark matter and the dark energy (the big crunch), waiting for the next explosion after several billions of years.

This answers one of the important questions, Who created this first cosmic egg to start the big bang? After all, something cannot come from nothing. If the answer is God, the same question is deflected to that of, Who created God? So basically, it looks like the cosmic egg and God are one and the same. Perhaps the religions are so confused and divided to recognise the true, visible, single God—the Nature—and are still seeking!

The Big Bang Theory in More Detail

The explosion produced enormous energy in the form of heat (billions of degrees kelvin) and matter in the form of quarks of subatomic particles. The heat represented all the energy of the entire universe in four identifiable forces. These are (1) the strongest nuclear force, (2) the stronger the electromagnetic force, (3) the weak nuclear force from the decay of atoms, and (4) the weakest gravitational force.

Then cooling of the universe occurred gradually with time and expansion. Subatomic particles from the quarks were scattered around as a primordial soup at very high speed, causing the newly formed universe to expand more and more with time.

When the temperature got cool enough, each of the protons and neutrons fused in pairs to form the simplest atomic nuclei (1p + 1n). Then, with more expansion and cooling, the electrons were able to fuse with the nucleus to form the simplest hydrogen atoms (1p +1n +1e; about 74 per cent still found in the universe). Some of the hydrogen atoms also fuse each other to form the helium gas (about 25 per cent still found in the universe).

At very high temperatures and in strong radiation, further fusion to create higher atomic-numbered elements is not possible. Within three seconds, the temperature dropped to one billion degrees. After 300,000 years, the universe cooled to about 3,000 degrees kelvin. Thereafter, the hydrogen and the helium gases clumped together in groups by gravitational force, and many billions of stars were formed. Gravity again grouped the stars together, and the galaxies were formed.

The Big Band and Events After

EVENTS AFTER THE BIG BANG

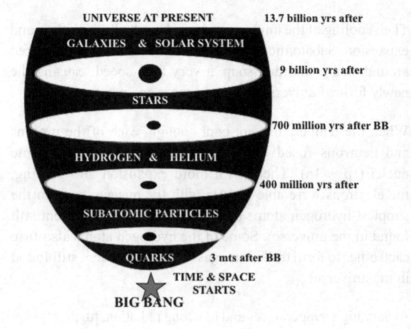

UNIVERSE AT PRESENT   13.7 billion yrs after

GALAXIES & SOLAR SYSTEM

9 billion yrs after

STARS

700 million yrs after BB

HYDROGEN & HELIUM

400 million yrs after

SUBATOMIC PARTICLES

QUARKS   3 mts after BB

TIME & SPACE
STARTS
BIG BANG

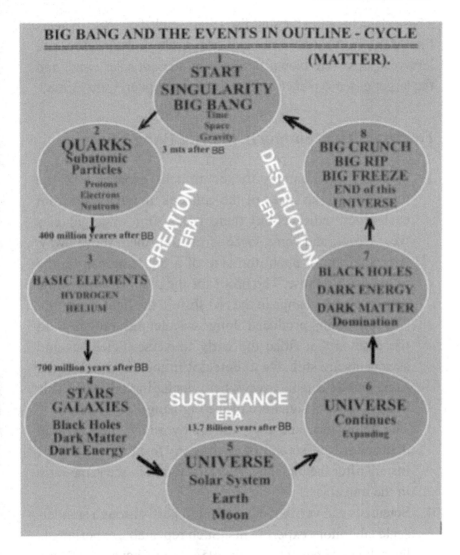

BIG BANG AND THE EVENTS IN OUTLINE - CYCLE (MATTER).

1
START
SINGULARITY
BIG BANG
Time
Space
Gravity
3 mts after BB

2
QUARKS
Subatomic
Particles
Protons
Electrons
Neutrons

400 million years after BB

3
BASIC ELEMENTS
HYDROGEN
HELIUM

700 million years after BB

4
STARS
GALAXIES
Black Holes
Dark Matter
Dark Energy

CREATION ERA

DESTRUCTION ERA

SUSTENANCE
ERA
13.7 Billion years after BB

5
UNIVERSE
Solar System
Earth
Moon

6
UNIVERSE
Continues
Expanding

7
BLACK HOLES
DARK ENERGY
DARK MATTER
Domination

8
BIG CRUNCH
BIG RIP
BIG FREEZE
END of this
UNIVERSE

Then our sun was formed from the solar nebula in our Milky Way Galaxy. From the leftover gases and dust in the disc of our sun, the planets formed by accretion process. Our sun is called a second-generation smaller star, meaning a new star formed from a dying first-generation larger star. The higher atomic-numbered elements were formed on the outside edges of the solar nebula.

These elements from the sun disc finally ended on rocky planets like Mercury, Venus, Earth, and Mars. The lighter debris and leftover gases were blown far away by the hot solar wind, and the larger gaseous planets, like Saturn and Jupiter, were formed.

Evidence That Supports the Big Bang Theory

1. Hubble's Law, namely the continuous expansion of the universe and the speed of the galaxies moving away from each other, indicates that there was a time when galaxies were closer, and some force is making them move away. An explosion of about the size of a big bang can be the only possible cause. The proof for the moving away of the galaxies is the change in the red shift in the light spectrum.
2. The big bang produced long wavelength radiation in the cosmos that filled the early universe. Scientists and astronauts are still able to detect it in the cosmos. It is the same and detectable everywhere, including on earth. This is accepted as a signature of the big bang.
3. The abundance and the ratio of hydrogen, helium, and deuterium (an isotope of hydrogen) formed in the first instant after the big bang. It is still found in the same ratio in the universe.
4. Scientists can verify the big bang in parts through smaller-scale laboratory experiments. Such repeatable experiments have been conducted since 1954. They have tested and proved the theory to be true.
5. As mentioned previously, related experiments using the Large Hadron Collider are still being performed at the CERN laboratory (from twenty-one countries, including Europe, America, and Japan). The underground laboratory

is twenty-seven kilometres long and located in Geneva on the Swiss-France border.

6. Darwin's natural selection and evolution theory and the big bang theory complement each other in respect to the enormous time period and the slow, gradual changes based on the cause-and-effect principle. Both theories demand many millions of years for a gradual evolution. A coordinated and parallel flow of development is noticeable between both living and non-living things (matter) in the history of the universe. Changes are inevitable in both theories, and all events happen with reasons other than a few accidental incidents.

7. Of all the theories, the big bang theory is the only sensible concept that can answer all the questions about creation and the Creator. It gives believable explanations to the events about the past, present, and the future of our universe, including the conditions of life on our planet.

The Role of Gravity in the Events of the Universe

Gravity is a force that pulls cosmic particles and objects towards each other, keeping them bound in clusters without straying loose in the vast cosmos. Anything that has a mass and rotates on its own axis generates a centrifugal force—a whirlpool—towards its centre.

Gravity is an entropic force produced from the beginning of the universe, when particles were tossed out into space after the big bang. The sudden force created from the bang initiated the spin, and vibration on the subatomic particles continued, which helps clumping. The acquired electromagnetic force in the particles can further enhance this spin, and thus collectively, a resultant gravity persists. Dark matter and the dark energy could also be involved. Gravity can only pull, not repel like electromagnetism.

The heavier and closer the object is, the stronger the pull. The gravity of earth keeps and holds us to the ground. It also keeps the moon orbiting around the earth. Likewise, the gravity of the sun keeps all its planets rotating in their orbits. In a similar way, our sun (a star) orbits around the galactic centre, and the galaxies orbit around the clusters, and so on. Gravity maintains and provides order to the universe.

Galileo was the first to discover the basics about gravity by dropping balls from higher positions. Newton went further, saying that gravity is what keeps the planets rotating in a circle around the sun. He derived a mathematical formula—$F = G \times Mm/R^2$—where F is the gravitational force, G is the gravitational constant (6.67×10^{-11} in SI units), and R is the distance. According to this law, an object twice as far from earth exerts a quarter of the gravitational force than one closer.

Einstein, in 1915, went into details via his relativity theory. He defined gravity as an effect, not a force. Namely, he said that when a free-falling object movies through the space-time curve, it also bends its path and predicts the phenomenon of gravitational waves. This was confirmed a hundred years later, in February 2016, when two black holes collided, producing huge gravitational waves about 1.3 billion light years away from us.

Gravity is the weakest of the four fundamental forces that erupted soon after the big bang. It could be the initial explosion that caused the quantum particles to whirl or spin and remain as the continuing gravitational force in all the cosmic objects formed by the same original particles. The gravity plays its most important role in creation and in the development of the events in the universe.

Gravity causes the formation of stars, solar system, planets, and moons. It binds heavy objects in space, like galaxies, stars, planets, dark matter, and black holes. It also holds the stars together within the galaxies. In addition, it keeps the moons bound to the planets.

Objects with a larger mass have larger gravity. An object that weighs 150 pounds on earth will weigh 354 pounds on the Jupiter, which has more gravity. When two objects of the same size are dropped, the heavier one will hit the ground first.

Gravity can be overpowered by speed if ejected upwards. To escape the gravity of earth, an object must travel upwards at a minimum speed of seven miles (11.4 km) per second. To do this from the sun, the speed needs to be 384 miles per second.

Lighter gases, like hydrogen and helium, can easily escape the surface of earth's gravity. But heavier oxygen molecules are retained in the earth's atmosphere by gravity. Presence of oxygen is one of the reasons life survives on earth compared to other planets.

It has been found that time runs slower at the surface of earth than higher above due to the difference in the gravitational force.

We have seen the moon pulling the oceans of the earth and the earth pulling the moon. Gravity is the reason for high tides during a full and new moon around the equator. A tidal force is actually the difference between the strengths of the gravity at two locations on the earth.

The Roles of Supercomputers and Advanced Telescopes in Research

Supercomputers

Supercomputers are so powerful that they can provide researchers with insight into phenomena that are too small, too big, too fast, or too slow to be observed in laboratories. Scientists also use supercomputers as "time machines" to explore the past and the future of the universe—up to billions of years—using today's known data. For example, in the year 2000, a supercomputer simulation was created that outlined a collision of our Milky Way Galaxy and the neighbouring Andromeda Galaxy in another 3 billion years! This simulation was performed on a parallel computer called Blue Horizon in San Diego. This could also tell us what will happen to millions of stars when these two galaxies collide.

Supercomputers—along with modern powerful telescopes on the ground, in space, and in large observatories—have helped researchers to run experiments and gather information. This was not possible in the days of simple telescopes and poorly equipped laboratories.

Researchers at the Texas Advanced Computing Center (TACC) at the University of Texas in Austin have used supercomputers to simulate the formation of the first galaxy. The scientists at NASA's Ames Research Center in California simulated the creation of stars from cosmic dust and gas. Supercomputer simulations also made it possible for physicists to answer questions about the future unseen universe. These supercomputers have even discovered that the distribution of the invisible dark matter is about 25 per cent and the dark energy makes up more than 70 per cent in the universe respectively.

Modelling based on supercomputer simulations is also used to help researchers understand earthquakes. Supercomputers also use them to help formulate advice on mining and oil explorations. In the medical field, they help in unravelling the mysteries of protein folding, mapping the bloodstream, and in modelling viruses (swine flu virus, for example) for vaccine development and possible mutations in viruses.

Advanced Telescopes

Advanced telescopes are developed periodically for usage on the ground and in space, within and outside the earth's atmosphere (for example, on the space stations). Starting from the first Galileo simple refracting type of telescope in 1609, telescopes were improved greatly to include reflecting types and radio telescopes that use radio waves from distant celestial objects to create images.

More recently, we have developed the ability to use X-rays to create images from objects like the sun, stars, and supernovas as they emit lot of these radiations. This includes using gamma rays to study events happening in the black holes, supernovas, and pulsars.

The Hubble Space Telescope was the earliest advanced telescope to orbit the earth, launched by NASA in 1990. By the year 2000, the United States, Japan, and Germany, in collaboration, built the Large Binocular Telescope, stationed in Mount Graham, Arizona.

Then in 2003, the Spitzer Space Telescope was launched. It follows the earth going around the sun, collecting various data.

Then came the Fermi Gamma Ray Space Telescope in 2008, built by NASA, Japan, and five other countries in Europe. It can measure the powerful radiation of our universe and watch the massive black holes.

One year later, NASA launched the Kepler Mission Telescope as a planet hunter in the outer orbit, about 46,478,000 miles in space.

In 2013, NASA, Canada, and European countries launched the James Webb Space Telescope.

Then, construction began on the Giant Magellan Telescope by a consortium of nine universities and researchers. The land-based telescope is stationed in a Chilean observatory.

Finally, the Large Thirty Meter Telescope is in the design stage. It will be able to look back to the formation days of the galaxies and stars.

Looking at the fast developments of the telescopes, we have to accept that science has studied and observed the creation process while the past revelations were based on guesses.

The Creation of Matter

As mentioned earlier, the creation started soon after the explosion with the quarks of subatomic particles and the emergence of the four fundamental forces. Expansion followed because of the need for space. Cooling occurred with expansion and time. The subatomic particles in the quarks soon underwent nuclear fusion and formed the simplest nuclei. After more cooling took place, the negatively charged electrons combined with almost all the unstable positively charged nuclei (1p + 1n) to form the basic neutral hydrogen (1p + 1n + 1e) atom with one proton, one neutron, and one electron. This occurred about 400,000 years after the big bang, when the temperature dropped to about 4000 degrees kelvin, at which point the electrons were able to bind with protons.

Until about 700,000 years after the big bang, the universe was dominated by severe infrared and ultraviolet radiation.

Next, helium was formed when two protons, two neutrons, and two electrons combined (2p + 2n + 2e). This gas is second in abundance in the universe at 24 per cent, compared to hydrogen at 75 per cent. This percentage (ratio) of these two dominant elementary gases are still constant in the cosmos.

At this stage, the universe was dark, mostly filled with hydrogen gas, and was glowing with leftover radiation from the big bang.

Next, the stars were formed in the universe when gravity slowly clumped the hydrogen gas and helium into dense groups, and the burning stars illuminated the universe. More and more large volumes of helium and lithium were formed within the burning stars. Observations of the background cosmic microwave radiation were used to date the ages of the stars.

Whenever matter is formed, antimatter is also formed, but they did not interact.

Then billions of stars grouped together due to the gravity and formed billions of galaxies. A galaxy is a spiral, whirling disc full of stars on the arms, circling around a dark centre, which is a small black hole.

Billions of galaxies were towards each other by the gravitational attraction force. They collected together in clusters and superclusters to form the universe. The time taken for the galaxy formation is calculated to have occurred about 700 million years after the big bang.

In our Milky Way Galaxy, our sun (a second-generation star) and the solar system was formed about 4.6 billion years ago, about 9 billion years after the big bang. Our moon was formed about 4.45 billions of years ago, when a large asteroid the size of Planet Mars hit our earth. The split-off pieces and other debris clumped together by gravity and became our moon.

Our Milky Way Galaxy

If you go out on a clear night, away from city lights, and look up at the vast sky, you can see millions of stars. You can also see that these millions of stars are in a faint, foggy band encircling the sky. This band and the collection of stars is our Milky Way Galaxy, the galaxy in which we live. There are about 200 billion stars in our galaxy, and it is about 100,000 light years in diameter (1 light year = the distance travelled by light in 1 year = 9.5 trillion kilometres, or about 6 trillion miles). There are six spiral arms to our galaxy.

- Outer, or Cygnus
- Inner Perseus or local arm
- Orion arm, where our solar system is located
- Sagittarius arm
- Scutum-Crux arm
- Inner-most Norma arm, swirling out from the centre

OUR GALAXY- Diagrammatic image of the Arms.

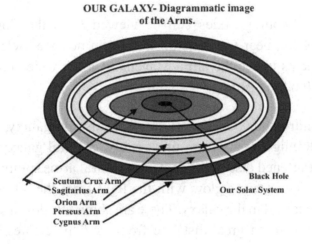

Scutum Crux Arm
Sagitarius Arm
Orion Arm
Perseus Arm
Cygnus Arm

Black Hole
Our Solar System

The central part of the galaxy contains about 90 per cent of all the stars (older stars, first generation). The third arm, the Orion, contains our sun and solar system (newer stars, second generation). Our sun is located in the outer disc in the Orion arm. The Orion circles the centre once every 230 million years.

New stars are born out of vast clouds of dust and gas located in the arms of the galaxy. There are about 200 billion stars of various sizes and stages of their lives.

Every large galaxy contains a massive black hole in the centre and a quasar (active galactic nuclei—AGN) brightly glowing powered by energy obtained from the black hole. The quasars are also called radio stars as they are a strong source of radio waves.

According to scientists, our sun was formed nearly 4.5 billion years ago and will live for a further 5 billion years before exploding into a red star, giving more dust. This red star will live for another 500 million years, until it finally shrinks in to a white dwarf star that will take billions of years to cool.

Our neighbour galaxies can be viewed from the Southern Hemisphere, near the equator. People do not need telescopic instruments to see the large Magellan Cloud, which contains about 10 billion stars.

The Andromeda Galaxy is another distant galaxy, about 2,300,000 light years from us. It is also a spiral galaxy made with a flattened disc at the centre and spiral arms spinning out from it. These arms glow with the light from the 300 billion stars contained in the galaxy. There are about two dozen smaller galaxies found a great distance from us. This creates a high possibility of similar solar systems, probably with liveable, earth-like planets.

By passing galactic light through prisms, astronomers have confirmed the movement and continued expansion of the universe. The light waves were seen stretched, indicating that there is motion away from us at a much faster rate. The more distant a galaxy, the faster it recedes.

Formation of Our Solar System, Planets, and Moon

Approximately 4.5 billion years ago, our solar system was just a cloud of dust and gases known as the solar nebula (burning star) in our galaxy. Gravitational spin collapsed the materials to form our sun and its surrounding accretion disc at the centre of the nebula. Thus, our sun is a second-generation star. Our solar system incorporates matter created by previous generations of stars.

The debris of the solar nebula disc became the planets through a process called accretion. This disc was composed mainly of hydrogen and helium, with other elements in smaller proportions. From the spinning of nebula disc, the debris and particles dispersed and started clumping together due to gravitational pull. Over a few millions of years, these chunks of particles continued to coalesce into planets, asteroids, and comets. The inner, smaller planets closer to the sun—Mercury, Venus, Earth, and Mars—are rocky with heavy molten mineral elements from the burning sun (about 2000° kelvin).

At the same time, the solar wind pushed far away the leftover lighter elements and the gases, such as hydrogen and helium, and formed the giant gas planets such as Saturn and Jupiter. Thus, these larger planets are more gaseous and cooler, about 50° kelvin).

The planets inherited their motion from their original disc rotation around the sun. They continue to orbit in the same direction and in roughly the same flat plane but at different speeds according to the size, distance, and contents. Earth's rocky core formed first with molten heavy elements. The lighter materials settled to form the crust. The flow of the mantle beneath the crust causes plate tectonics, the movement of large plates of rock on the surface of earth.

LIFE CYCLE OF THE STARS

The heat that was trapped deep in the core bursts out as volcanoes that develop the mountains, hills, and valleys. The various gases produced after volcanic eruptions are pushed into the earth's atmosphere. Some fall back to the ground as acid rain, which again reacts with minerals on the earth to create more chemical compounds.

At the beginning, all the present continents of our planet were together and closer to each other as one rocky mass surrounded by the ocean. But with time and natural events—like volcanoes, earthquakes, and tsunamis—the hot rocky mass got separated or joined and the ocean water rushed into the spaces between the masses to form various oceans. In this way some land masses went under the sea while some areas that were under the ocean got exposed as new land areas.

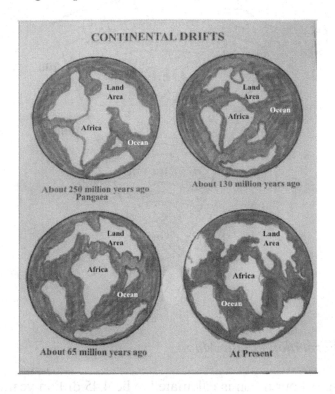

This occurred very slowly over billions of years. Supercomputers using geological data have calculated this history.

Our solar system, including the earth, is expected to vanish in about few billions of years. But before this, life on earth could perish due to the gradual increasing of environmental heat from the sun. The heat can cause the evaporation of the earth's water, and it will also become too hot for life to survive.

EARTH LAYERS IN SECTION

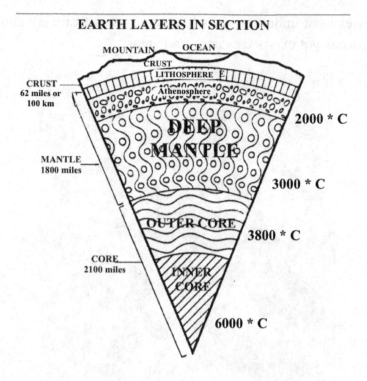

The Formation of Our Moon

The Formation of Our Moon

The age of our moon is calculated to be 4.45 dillion years old. During the early days, when our planet earth was formed, there were too many cosmic rocks around. One large asteroid impacted our planet, chipping out a big chunk. This chipped part of our planet, along with other debris floating and the large object that crashed, coalesced due to gravity. This resulted in the formation of our moon. We call it moon because it was formed of our planet, it stays close to earth, and it continues to rotate our planet. The other independent planets rotate around the sun because they were formed from the sun. The lost mass on the top side of our planet it what gave earth its tilt.

While dust and debris form moon, when icy bodies collide into earth, they deposit water on the planet's surface, around 4,280 million years ago, which is the key factor for life to develop.

Origin of Elements Summary

1. Hydrogen is the simplest and the first element formed after the big bang, when the first nucleus fused with the first electron.
2. Most of the helium formed later in the stars, when two hydrogen atoms fused. Helium and hydrogen make up to 99 per cent of known matter in the universe.
3. Carbon, oxygen, and nitrogen were formed by the energy-releasing process of the nuclear fusion within the stars.
4. Elements lighter than iron (26) are created by a process known as stellar nucleosynthesis, which occurred deep inside the cores of the stars.
5. Elements heavier than iron (26) up to lead (82) were formed by consuming energy rather than releasing it. This occurred in stellar explosions called supernovae.
6. Elements heavier than lead (82) tend to fall apart and are called radioactive.
7. Gold, platinum, and uranium were produced by collisions between ultra-dense objects in the neutron stars. The neutron stars are produced when massive stars larger than our sun explode as supernovae, leaving behind super-dense magnetized balls composed mainly of neutrons.

Periodic Table of Elements Supporting the Natural Creation of Elements

Scientists arranged the elements in the ascending order of their atomic numbers. They found the elements falling in a definite pattern of groups in accordance to their specific physical and chemical properties. The gases, metals, non-metals, and halogens are in specific groups, columns, or rows.

The way the elements were created and their order in the periodic table is another good example of nature's role in the creation process. This also gives evidence to negate the spontaneous creation theory in regard to elements in the universe.

Hydrogen and helium were the first two elements, first formed by fusion at very high temperatures soon after the big bang. The further building up of elements was shut down immediately after about three minutes of the bang because the temperature cooled due to expansion. Nuclear reaction—fusion—can take place only at very high temperatures. Because of this, there were only three elements in the early universe: 74 per cent hydrogen (1p + 1n), 25 per cent helium (2p + 2n), and very little lithium (3p + 3n).

The hydrogen and the helium clumped together by gravity, and stars were formed about 380,000 years ago. The centre of the stars has a temperature of about 15 million degrees.

But when the stars continue to burn and are about to die (nebula), the temperature increases to about 100 million degrees kelvin. Such high temperatures are required for helium to fuse further and become carbon (6). When low-mass stars were dying, lithium (3), carbon (6), and nitrogen (7) were formed in the peripheries of the nebula.

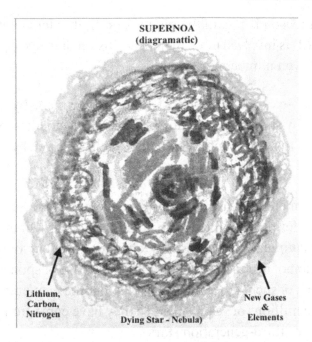

SUPERNOA
(diagramattic)

Lithium,
Carbon,
Nitrogen

New Gases
&
Elements

Dying Star - Nebula)

When massive stars (red giants) exploded, higher elements, from oxygen (8) to rubidium (37), were formed. Explosion of white dwarfs formed silicon (14) to zinc (30). Then most of the higher atomic numbered elements (41–92) were formed when neutron stars merged and exploded. The heavy, higher atomic numbered elements ended naturally on the rocky planets (earth), asteroids, and meteorites.

Iron is the most stable element in the universe as it is formed in the core of the massive stars having higher temperatures. When these massive stars die, the iron collapse and all other rare elements, like gold, are formed in the last two seconds. Element gold (79) is precious as it is rare. The even atomic numbered elements are about ten times more prevalent in the universe than the odd atomic numbered elements. For example, there is ten times more oxygen (6) than nitrogen (7) in the universe.

Natural reason being, in the giant stars (higher temperatures), helium (2) is involved more in nuclear fusion than hydrogen (1), and the even number gets added.

On earth, oxygen is the most common element by weight followed by silicon, aluminium, and iron. In our bodies, oxygen is also the most common element by weight followed by carbon, hydrogen, and nitrogen.

The origin and the distribution of the elements are also added evidence for nature's role in the creation process. On the other hand, manipulating such minor changes to occur on a tiny planet cannot be practically possible in the case of spontaneous creation. As such, it becomes obvious that the elements in our body are millions of years old since they have come from first-, second-, or third-generation stars.

Much larger atomic numbered elements (95–118) are made synthetically in the laboratory by bombarding smaller elements through performing a fusion reaction.

Changes Occurring with the Earth and the Moon

Interestingly, during the development of the universe, the duration of a day on earth and the distance of the moon from the earth have been increasing slowly over time. The reason is obviously nature and cannot be God.

- About 900 million years ago, a day on earth was about eighteen hours long, and the moon was about 350,000 kilometres from earth.

- About 650 million years ago, the day increased to 20.7 hours, and the distance to the moon increased to 357,000 kilometres.
- About 360 million years ago, a day was up to 21.8 hours. The temperature dropped from 34 degrees Celsius (93 Fahrenheit) to 26 degrees Celsius (78 Fahrenheit).
- Since 45 million years ago, the earth's day increased to twenty-four hours, and the distance to the moon is 378,000 kilometres.

If God created the universe, why would these changes occur incrementally? Surely an all-knowing God need not change these. Rather, God would create a universe with permanent optimal conditions from the very start.

Earth Is Fine-Tuned to Have Life

The universe was created with all its natural forces and respective natural laws. The solar system is ideally grouped, with the earth suitable for the development and existence of living things. The features of our earth are naturally fine-tuned for life to survive, and it is the only known planet in our solar system that supports life. So far, no other planet has been discovered with the precision to fill the requirements for the living creatures.

(1) The size of earth is what makes it first and foremost suitable for life. If the size of earth was slightly larger, its gravity would have been too strong for the lighter hydrogen and helium gases to be pulled on to the earth's surface. This would have caused the wrong mixture of gases (creating a hydrogen atmosphere)

on our planet, making it impossible or difficult for living organisms to survive.

And on the contrary, if the size of earth was little smaller, the life-sustaining oxygen, gas, and even water molecules would have escaped the earth's surface. Again, no life would be possible on earth. But earth's optimum size is able to hold slightly heavier gases like oxygen, nitrogen, and carbon dioxide on the surface for the metabolism of the animals and plants.

(2) The ideal positioning of earth in our solar system allows it to maintain life. The distance of our planet from the sun is optimal, with living creatures receiving the correct amount of heat and energy. If earth was moved about 5 per cent closer to the sun, there would be overheating, and all life would have perished about 4,000 million years ago. On the other hand, if earth would have positioned just 1 per cent further from the sun, heavy covers of snow would have formed all over the surface about 2,000 million years back, and no living organisms could have thrived.

(3) Another important factor is that Earth takes twenty-four hours—a twelve-hour day and a twelve-hour night—for a full rotation on its axis. This gives us the equal day and night with optimum temperatures and energy from sunlight. This also gives us the balanced active day and passive night as resting periods. The 365 days to go around the sun gives us four equal seasons. In comparison, planets like Venus take 243 days to go around the sun and has longer light periods and shorter dark nights, making it unsuitable for life.

(4) The pathway that the earth takes around the sun is more circular compared to the elliptic pathway taken by other planets.

If earth also goes on elliptic path, it would be far from the sun during some months and very close for many months from the sun than the present moderate distance throughout the 365 days. Again, this will result in life not being possible to thrive. These fine conditions are expected to go on for few more millions of years.

(5) The location of our solar system at about the outer edge of our galaxy (Orion arm) is also an ideal location. Near this location, there are not many stars around to interfere with our planet regarding electromagnetism, rotations, and radiations from neighbours. Fortunately, there are few or no disturbances to us from the electromagnetic forces and radiations from other cosmic objects.

(6) Finally, the earth's 23.5° vertical tilt around the North Pole towards the sun provides the various four seasons (four shorter seasons in the Northern and Southern Hemispheres, with two longer seasons near the equator. As mentioned previously, the tilt occurred accidently about 4.45 billion years ago, when a heavy cosmic object about the size of Mars crashed on earth.

Chapter 3
Life on Earth

The Origin of Life on Earth

Let's begin with a brief review. Earth came into existence from our solar system about 4.57 billion years ago as a hot, hard, rocky planet. At the beginning, all present-day continents were just a single mass. Then the first few continents formed closer to each other, and oceans developed between them.

It took about one billion years for water to appear on our planet, when icy cosmic objects hit our planet, and the volcanoes erupted. Water gradually surrounded the land mass as oceans when the surface cooled. With volcanoes and earthquakes, molten rocks formed into mountains and valleys. Ocean water got into the lower land areas, separating the land mass and forming continents.

The volcanic eruptions also brought hydrogen, nitrogen, water vapour, carbon dioxide, ammonia, and some hydrogen sulphide gases to the earth's atmosphere. Hydrogen, being light, escaped from the earth's atmosphere, while water vapour condensed to water as rain. These simpler inorganic gases in the presence of water vapour in an environment of lightning and thunder combined to form less-complex organic compounds that were the precursors of living organisms.

From simple inorganic gases like water vapour, carbon dioxide, ammonia, and methane, organic compounds like basic sugars, amino acids, and fatty acids were formed. These simple organic compounds combined, and non-cellular life structures began. As time passed, these organic masses continued to develop into non-cellular organisms (virus-like) showing life qualities. With more time, these structures evolved into complex cellular organisms with a cell wall (bacteria-like), nucleus, and other organelles. The single-celled organisms gradually evolved to multicellular, and after many millions of years, to higher-level species, the latest being the human species at the peak of the evolution. It is important to note that the earth had a prehuman era of about four billion years.

Among the living species, competition and struggle for existence arose because of environmental conditions and predators. It led to the survival of the fittest while the others perished or were left out of the evolution process. Mutations helped survival, and newer species evolved. As a result, we now have billions of plant and animal species that have evolved to adapt over a long period of time.

In the history of our planet, there had been about five global extinctions, leading to over 75 per cent of all the species erased. This was due to drastic climatic conditions, mostly resulting from long, complete snow cover; impact and explosions of other cosmic objects, like comets, meteorites, and asteroids. The heat and dust that followed these impacts covered the entire atmosphere of earth for long periods of time, suffocating life.

The notable extinction was the dinosaurs during the prehuman era. Dinosaurs dominated earth about 125 million of years ago but were completely eradicated by about 65 million of years ago as a result of a large meteor that impacted near Mexico.

The Biochemistry of the Origin of Life (Brief)

- By 2,200 million years ago, the aerobic microbial organisms with mitochondria developed. The environment of the primitive earth developed to allow simple chemical reactions that synthesised organic compounds from basic inorganic precursor elements.

- Primitive earth's atmosphere contained water vapour, methane, ammonia, and hydrogen. The warmer environment from the volcanic eruptions and lightning (electric), together with thunder shocks, promoted these chemical reactions to occur in a warm, wet medium (water vapour) on the surface of our planet and under the shallow sea bed.

- Early life on earth was formed through a series of chemical reactions that made simple, primitive, inorganic, free atoms such as hydrogen (H), oxygen (O), carbon(C), and nitrogen (N) into inorganic molecules such as hydrogen (H_2), water vapour (H_2O), carbon dioxide (CO_2), methane (CH_4), and ammonia (NH_3).

- Then methane and water vapour combined to form simple sugars, fatty acids, and glycerol. Methane, water vapour, and ammonia combined to form the amino acids.

- Simple sugars joined through polymerization to form the polysaccharides. The lipids formed from simpler fatty acids and glycerol. Proteins, nucleotides, and nucleic acids were formed by the condensation reaction of amino acids.

- The larger organic molecules (probiotics) that grow and divide were under the control of the nucleic acids, RNA and DNA.

- Next, proteins and lipids combined to form the membranous lipoproteins that enclosed the other complex organic compounds to become a primitive cell. The nucleic acids controlled these functions. This structure, called a cell, can perform many functions of life.

Laboratory Evidence to Support the Conversion of Inorganic to Organic Elements

In 1952, Millar and Urey at the University of Chicago demonstrated some experiments to prove these natural reactions. They showed that from simple inorganic molecules—like water, ammonia, and methane—could be converted to smaller organic molecules like amino acids in the laboratory given the similar atmospheric conditions of our primitive earth. After about a week of the experiment, they found that about 10 to 15 per cent of the carbon was converted into organic compounds, including about 2 per cent in the form of an essential amino acid of life, glycine.

They also demonstrated in another experiment that one of life's building blocks, the nucleotide base adenine (one of the four bases of RNA and DNA), could be produced from hydrogen cyanide (HCN) and ammonia (NH3) in water under similar conditions. The cells slowly developed to respond to the environment, metabolized, and reproduced like a primitive bacterium.

Chapter 4
Evolution of Life on Earth

Natural Selection Theory of Darwin

Understanding the history of life on earth requires a grasp of the depth of time and the breadth of space. We must always keep in mind that the time involved in this history is very long, many billions of years, compared to the life span of the human generation.

In 1831, Charles Darwin began a five-year voyage around the world to collect and study the biological evolution of plants and animals. This eventually led to his book, *On the Origin of Species.*

During his study, he discovered that there has been a competition among living organisms and struggle for existence. When competing, the fittest survived and continued to reproduce. This process is called natural selection. And eventually after a few generations, new species evolved.

Natural selection is a process that causes populations to change over time in a changing environment. This change of evolution happens gradually, over a very long period, from tens of thousands to many millions of years. Speciation happens when a species splits into two or more branches of improved species. But all species come from a common ancestor.

Evolution of the Animal Kingdom

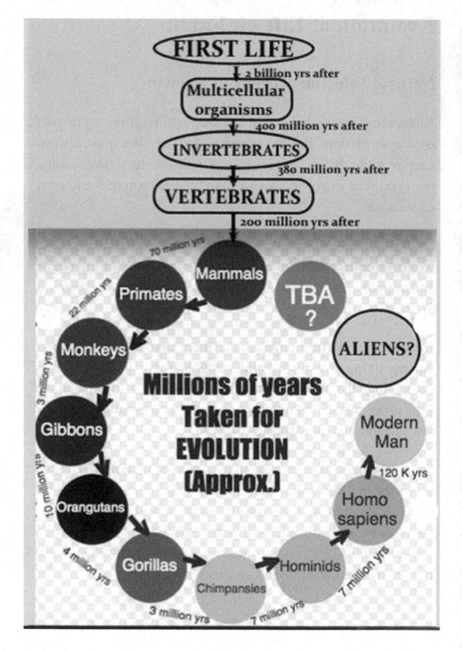

Evolution of the Plant Kingdom

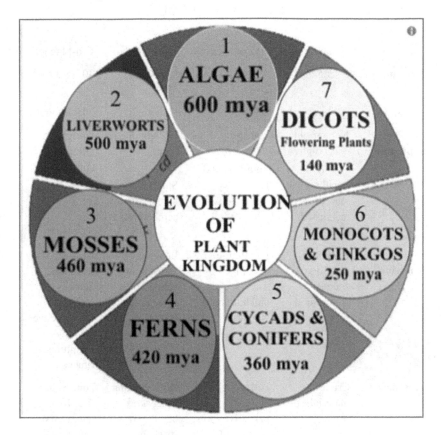

EARLY HUMAN CHARATERS IN
EVOLUTION - TABLE.

	Homo habillis	Homo erectus	Neanderthal	Cro-Magnon 1
Time period	3 – 1.5 mya	1.5 mya – 200 Kya	230 Kya- 35 Kya	40 – 8 Kya
Life span	< 20yrs	<20 yrs	< 30 yrs	30 – 40 yrs
Physical appearance	Not upright 3.5 ft high. Small frontal lobe, thick hair, black skin.	Upright Small frontal lobe.	Upright, 5.5 ft high. Small frontal lobe, Large cerebellum.	Upright, taller, less strong, like modern man.
Social behavior	Language primitive sounds, gather food, scavenging.	Languag-e primitive sounds, gather and hunt food, Nomads.	Language by signs, hunts, cooperative, burials, caring disables.	Language sentences, hunts cooperative, burials, Cave paintings, sculptures, rituals.
Terminology	Crude materials.	Hides for clothing, primitive tools, rocks tied to sticks.	Sewn clothing using skins, building with sticks for shelter, spears.	Pottery, use fire, Tools u bones carved, chisel, fish hooks.

In nature, animals and plants have the tendency to overpopulate their species in order to propagate and have a better chance of survival. When there is overproduction, variations of traits resulting in character, morphology, and behaviour (phenotypes) occur. With these differences, there will naturally be competition among the same species for water, food, and space. So there will be a struggle for existence. Hence, adaptations and developments are necessary to survive. For example, the predators must run faster and catch enough prey. On the other hand, the prey have to camouflage themselves or run faster to escape. Therefore, the fittest continues, mostly with better adaptations. Those that survive continue to reproduce and will eventually become the most common in the population. The genes and the order of the makeup (genotype) will also change slowly within a population of organism. The inherited traits are passed on to future generations, and a new species emerges through an evolution.

It should be noted that this process of natural selection is an evolution of the population, not of an individual. This evolution is specific for that particular environment and may not be repeatable in a different environment.

Origin of life on Earth
(million years ago)

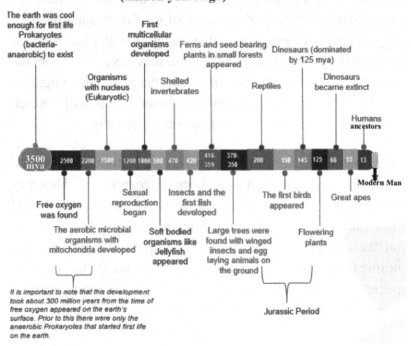

Evidence That Supports the Natural Selection Theory of Darwin

Fossils

Fossils are formed when organisms die and are buried in sediment, allowing little decomposition of the organisms. Over time, calcium in the bones and other hardy tissues are replaced by minerals, and then the sediment gets changed into rocks. As time progresses, various sediments are deposited in layers with other fossils over the previous one. So the oldest fossil will be on the bottom, and the latest on the top. From the appearance, the abundance, and the type of fossil in each of the layers, we can understand the progression of the species that lived in that location over time. The following table gives the time periods and the locations where our ancestors lived in the order of evolution. It is not easy to locate and excavate all the fossils underneath as the geography of our planet is also changing with time. Some of the fossils are still unreachable because they have gone under the ocean and vice versa. At the same time, the population migrated too! As such, there are some "missing links" yet to be discovered to connect the chain of physical evidence perfectly.

For example, the missing link between the Homo erectus and Homo neanderthalensis has not been discovered yet. This may be because of the migration that occurred during this period from East African Rift Valley north and east of the globe. But there is other evidence from comparative anatomy and rudimentary organs to support the changes in the evolved species. Similarly, the missing link between the aquatic to terrestrial life is also not clear due to changes in habitat.

History and Geography of Hominid Fossils

HOMINID FOSSILS	(PLACES AND TIME)	
Genus and Species	**Location of Fossil**	**Estimated Age of Fossil**
1 Ardipithecus ramidus	Aramis in Ethiopia	4.4 Million years ago
2 Australopithecus anamensis	Knapoi in Kenya	4.0 Million years ago
3 Australipithecus afarensis	Hadar in Ethiopia	3.2 Million years ago
4 Australopithecus africanus	South Africa	2 -3 Million years ago
5 Australopithecus aethiopicus	West Turkana in Kenya	2.5 Million years ago
6 Australopithecus boisei	Lake Turkana in Kenya	1.8 - 1.4 Million years ago
7 Homo habillis	Olduvai gorge in Tanzani	1.8 years ago
8 Homo erectus	Java in Indonesia	1.5 Million years ago
9 Homo ergaster	Koobi in Kenya	1.7 Million years ago
10 Homo heidelbergensis	Heidelberg in Germany	400K to 700 K years ago
11 Homo neanderthalensis	La-Chapelle in France	50,000 years ago
12 Homo sapiens	Cro-magnon in France	30,000 years ago.

Fossils can be dated to establish their approximate ages and hence, the period in which the organisms existed. Numerous extinct species are found as fossils. Modern species can be traced through fossil relatives to their distant origins.

A good example is the fossil evidence showing how the modern-day horse evolved through a period of 60 million years. Originally a 0.4-metre-tall, dog-like creature that lived in the rainforests, has evolved now to a 2-metre tall animal. In this process, a multi-toed foot evolved into a single-toed (comparative anatomy of hoof) that is better-suited for running fast.

Fossil fuels are buried organic materials formed by plants and animals that have decayed and been converted to combustible deposits such as crude oils, coal, and natural gas in the earth's crust over 650 million years ago.

*Anatomy and Vestigial **Organs***

Studies in comparative anatomy (bones and muscles) of the animals have shown that there is a common design from which these structures evolved to suit the requirements of the environments and conditions in which they live. The variations are shown in the animals' size, feeding habits, and housing types. For example, the hind thigh bones are displayed below to show how the variations have come from a common design.

Insert illustration of thigh bones

Vestigial Organs

Humans have many rudimentary organs, including the appendix, caudal vertebrae, pointed canine teeth, and small pieces of nictitating membrane in the eyes. These organs are not used, but they remain in our bodies as vestigial organs.

With evolution and the change in our lifestyles, these unused organs have gradually become rudimentary or obsolete. They were functional and essential for our ancestors. When these organs remain as evidence of our evolutionary history, it negates the spontaneous creation theory!

Vestigial Organs i Humans

- Nictitating membranes in the eye

- Vermiform appendix
- Caudal vertebrae
- Pointed canine teeth
- Muscles to move ears
- Few hairs on body.

Embryology

Comparative studies of developing embryos has proved that there is common basic anatomical design (tetrapods, tail, eyes, etc.) during the early stages of development. The similarities indicate that at some stage, all have come from a common ancestor—a common design—and deviated to prove evolution. This too supports the evolution theory and negates the spontaneous theory of creation.

Chromosomes, DNA, and RNA

The DNA code itself is a homology that links all life on earth to a common ancestor. The DNA and inherited traits of humans and chimpanzees are about 98.7 per cent similar. The sharing of DNA in different species is important evidence for a common ancestor and the natural evolution theory.

Chromosomes and DNA in living organisms are the most complex molecules in the world. An average human has about 50 trillion cells, and each cell has forty-six chromosomes. DNA is inherited, along with chromosomes from parents, and reshuffled in each generation. The number of chromosomes in organisms vary in each species. Our close ancestors—monkeys, chimpanzees, orangutans, and apes—have forty-eight chromosomes. Horses have sixty-four and donkeys sixty-two. Despite a one-pair difference, these two are able to mate and get an offspring, a mule. So no wonder this may be applicable to apes and humans during the evolution period.

Dogs and the chickens have twenty-two; the housefly and a tomato have twelve. Bacteria have just a single chromosome, but a hermit crab has 254. A small butterfly has 380, and a fruit fly has just 8. So the quantity of chromosomes does not govern the quality of the species. It is the specific design and sequences of its components that give various characteristics to species. The random way in which the numbers of chromosomes are distributed in organisms does not suggest the involvement of an intelligent Creator (God).

The DNA is the instruction manual of an organism and is fixed onto the chromosomes in the form of genes. Chemically, a DNA molecule has three basic components: a sugar (deoxyribose), a phosphorous-containing group (phosphate radical), and the bases, or the four nucleotides (adenine, cytosine, thymine, and guanine). The probable helix formation and the crisscross combinations, along with varying orders of the four bases, result in several combinations and different characteristics in the species. In the 1960s, scientists found a nearly infinite number of codes exist when A-T and C-G are arranged in rows and various sequences on such a long chain.

DNA plays a very important part in solving paternity confusions and in criminology cases to confirm causes and rule out potential suspects. Inferences obtained from the DNA analysis further support the evolution theory and the type of connection between various species and families in the living things.

Second Creation

The ability to create life naturally via science research in vitro (laboratories) is supposed to be the second creation. Medical discoveries like cloning, organ transplants, test-tube babies, IVF, stem-cell treatments, and gene transplants do not support the grand design theory of the Creator who is the only one with the authority to create.

In 1996, Ian Wilmot and Keith Campbell, with their colleagues at Roslyn Institute, cloned a sheep called Dolly without the involvement of any male component. It was a result of genetic research and an asexual way of reproduction, the first in mammals. They took a cell from the mammary gland of an ewe and grew it in a culture in a lab. They then fused this cultured cell with an egg from another ewe to "construct" an embryo. After fusion, they transferred the egg into the womb of a surrogate mother, where it developed into the lamb Dolly.

These events have overturned one of the most important and deepest dogmas embedded in our religious history; namely, that only God can create life by his set-up. It also opens up many further possibilities in terms of what may be possible by humans based on nature.

Researchers continued further from Dolly, which was cloned but genetically unaltered. Again in 1997, they cloned Polly and obtained the genetically altered offspring by transforming it with upgraded traits. They improved the original genetic makeup of the animal by removing the unwanted genes and adding genes that could give better performance. For social reasons, this research had to stop as some feared that this type of engineering in the wrong hands on wrong subjects could result in harmful effects

Pon Satchi

to society as a whole. But if genetic engineering is used wisely, it can also lead to a much-improved human species developed for peace and unity. When genetic engineering is used in agriculture for higher yields and disease-resistant hybrids, the world can eradicate poverty and hard labour in no time.

Genes and Gene Modifications

Gene distributions are another good evidence to support the evolution theory via natural selection. Gene mutation is an important event in the evolution for the speciation.

Nowadays, genetic engineering has developed so much that finer details of the genes, their locations, and their actions are detected with the help of modern tools and research. Artificial breeding for hybrids provides quicker results and confirms the natural evolution process. Scientists are so advanced now to work within the microscopic nucleus of the cells and can decode the genetic codes. Using modern technologies and tools, scientists are now able to extract desired genes from one organism and transfer them into the gene code of another organism to obtain a desired quality.

By altering the genetic code of an organism in the laboratory, scientists are also able to alter the characteristics and behaviours of the recipient organism, resulting in partially or fully changed species.

If this technique is incorporated with cloning procedures and IVF techniques, scientists can create a completely new species. This is a sort of new way to create life—humans using nature without anyone from heaven! This development contradicts our ancient beliefs and religious dogma regarding the Creator.

Gene editing and the transfer of desired genes are widely used in agriculture, animal husbandry, medicine, nutrition, and disease prevention. For example, we see that scientists have created many genetically modified (GM) agricultural products as hybrids in corn, soybeans, cotton, canola, and so on for higher yields and to resist diseases.

In 2003, scientists were able to sequence and map the entire human genome and read the complete blueprint for creating a human being. Following this, in 2010–2011, using genetic sequencing and gene editing, they detected previously unknown diseases. In addition, it is now able to treat some diseases caused by mutated genes.

Artificial Manipulation of DNA

DNA is the blueprint for our life process. Each cell of our body contains the complete genetic code for our whole body. The sections of DNA called genes instruct all our physical features and other qualities of our body.

Gregory Mendel founded classical genetics in the 1860s. In the 1950s, James Watson and Francis Crick at Cambridge described the three-dimensional structure of the DNA and founded modern molecular biology, essentially the science of DNA and its products.

The study of modern genetics and its applications stands as an important witness to negate the requirement for a God to be involved in creation and evolution. People have been amazed by the anatomy and physiology of living organisms, especially in humans. Our bodies are wonderfully developed and finely tuned to tackle the environment. Those who still believe in God argue that this Creator must be endowed with super-intelligence to create such fussy variations in several species.

But when scientists researched and discovered the procedures of decoding, detaching, and transferring of individual genes, that concept was negated! This became possible because of increased knowledge in science and advances in technology and tools (computers and lab facilities). It is expected that very soon, humans will be able to create life artificially following the natural scientific way.

Genetic Engineering and Artificial Creation

Perhaps one of the most significant uses of genetic engineering involves improving human health, including chronic conditions.

The study and observations of some insects that became resistant to pesticides and some bacteria losing sensitivity to certain antibiotics further support the evolution theory.

In the medical field, in 1983, scientists used genetic engineering techniques to manufacture insulin using a bacterium. They identified and extracted the gene that is responsible for the insulin production in the human pancreas. They attached this gene to a rapidly multiplying bacterium called E-coli using plasmid as the carrier.

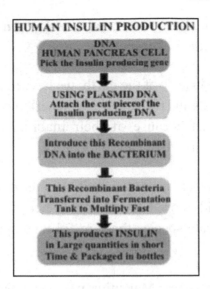

Since bacteria reproduce quickly, in a short time they were able produce lots of insulin when grown in a culture in the lab. After a few days of multiplying extracted the Humulin (insulin) from the E-coli, they marketed it to treat people with diabetes. Similarly, in 1990, biotechnology helped scientists tackle the bone marrow–related immune disease called SCID. Here the gene was extracted, introduced into the lymphocytes, and then infused into the bone marrow that produce the B and T cells that are responsible for immunity.

Recently, medical treatments are performed using stem cells. Stem cells are special cells with the amazing power to transform into any tissue or organ in your body. There are currently eighty to a hundred diseases that can be treated using stem-cell technology. Stem cells can be extracted from our bone marrow, blood, and from a foetus. Amniotic fluids are being frozen and banked to obtain stem cells for use when required. In fact, stem-cell treatment is a natural adaptation of the body, happening even before scientists discovered its use.

Periodic Extinction of Animals and Plants Due to the Environment and Predators

Throughout the history of the earth, we have noted that several species of animals and plants have been periodically eradicated, becoming extinct because of their inability to survive the competition, environment, and predators.

If God is responsible for creation, it begs the question why the Almighty would allow his creation to become extinct. Why wouldn't he create species that are fully equipped to survive throughout?

Birds like the dodo from Mauritius, passenger pigeons, laughing owls, Labrador ducks, Galapagos tortoises, Japanese sea lions, golden toads, even our ancestor Neanderthals and many more are all extinct now due to natural causes.

Global Extinctions Due to Drastic Climate Changes and Accidents

Periodic global extinctions and the restarts of life due to natural causes are further evidence that negate God as the Creator. Why would God create and not protect life that would become extinct?

There have been about five mass extinctions in the past that killed almost all (up to 90 per cent) of living organisms. The fossil records provide evidence of these mass extinctions and the restarts. The extinctions were caused by adverse climatic conditions (severe heat or an ice age) and by the impacts of large cosmic objects such as meteorites and asteroids on the earth's surface. The impacts produced severe heat and thick dust cover

over the whole atmosphere of the planet, which killed the animals and plants by blocking the sunlight and oxygen required for life.

Global Extinctions

Event	Date (Ma)	Description
Great Oxygenation Event	2,450	earliest oxygenation of atmosphere
Cambrian Explosion	542	rapid diversification of animal life
End-Ordovician extinction	446	86% of species lost
Late Devonian extinction	372	75% of species lost
End-Permian extinction	252	95% of species lost
End-Triassic extinction	201	80% of species lost
End-Cretaceous extinction	65	76% of species lost

To prove these dangerous impacts, on December 18, 2018, a meteor blast was observed over Russia's Bering Sea that packed ten times the power of the Hiroshima bomb! Fortunately, this exploded above twenty-five kilometres in the earth's atmosphere, and water below has neutralised without any damages on land or risk of another extinction.

1. The first extinction, of about 86 per cent of the living organisms, occurred during the Cambrian period (550–450 million years ago) due to catastrophic methane release from the sea.
2. The second extinction occurred around 375 million years ago, resulting in the extinction of 75 per cent of all living organisms due to volcanic eruptions, and emissions of carbon dioxide and sulphur dioxide gases that produced acid rain all over earth and flooding.
3. The third extinction was 251 million years ago, when about 95 per cent of living organisms became extinct due to severe climate change, resulting in the earth being covered with ice for long period (ice age).

4. Around 200 million years ago, 80 per cent of living organisms were eradicated due a huge meteorite impact that caused heat and dust cloud cover all over the earth's atmosphere.

5. By 65.5 million years ago, another meteorite impact eradicated about 76 per cent of life on earth. The impact struck what is now called the Yucatan Peninsula in Mexico. It eradicated the entire dinosaur population and another five billion species of life. The severe heat and dust from the heavy impact covered the entire atmosphere, and most life perished.

Global mass extinctions are at odds with the theory of an intelligent creator, the God. Why would God create, allow the nature to eradicate the entire work, and then recreate again and again? These events also point to the fact that nature is the ultimate master of the universe. Nature is unfolding as it should, and it is life that has to survive.

The universe is full of matter, energy, and forces in motion. Thus, natural causes that impact life and expose its fragility should be expected. Even today, despite our advancements in technology, natural disasters—such as floods, earthquakes, and tsunamis—have a way of creating devastation that no one can prevent. Nor does history suggest that the Creator has either created a creation to survive or attempted to prevent such devastation and destruction. Why create to have your creation destroyed by nature?

Periodically in life, diseases, predators, and genocides in the name of religion, colour, and ethnicity also ravage the world.

Critics of Darwin's Natural Selection Theory

Some people still believe in the spontaneous creation theory and are critics of Darwin's natural selection. The fundamental reason is due to their lack of understanding about the many millions of years of the history of earth and the periodic drift of the continents. In the history of our planet, the land mass and the oceans have been drifting widely with time. As such, the localities for evidence went towards underwater and new land areas. In addition, there were two mass migrations of our ancestors (hominids) from the East African Valley towards the north, east, and west. It is understandable that the people who believe in spontaneous creation are also unable to compute the huge time (millions of years back) involved in this issue. They are taught to think in very short periods of time, such as ten to fifteen years per the religious stories they believe.

Without many tools, like calculators and computers, Darwin in 1831 was able to compute and grasp the true history of our planet and the life that existed for many millions of years. He was so clever to discover a theory for the evolving living things in various locations that comprehended this length of time. The big bang theory, after centuries, complements his long-time factor for a gradual progress in a cause-and-effect system. Religions, on the other hand, assume that humans are God's preferred species and have come out with the spontaneous creation theory that is limited and deals in very short time frames.

Migration of our ancestors from the East African Valley has spread the fossils of the human species all over the planet. Thus, a 100 per cent percent chain is not yet completed.

Another reason for some to disagree with Darwin's theory surrounds undiscovered fossils, especially the few missing links. As outlined above, such reasoning stems from a lack of understanding of the vastness of the areas on the continent and the ocean drifts that have occurred periodically. The changing land areas and the ocean makes it impossible to excavate everywhere to find all the missing links. Drilling for fossil fuels in the desert and in the sea where there are no or few vegetation now is a good indication that more fossils are still hidden and yet to be revealed.

Migration of Our Ancestors: The Human Story Starts in the East African Valley

The origin of our species is in Africa, where we started to evolve from the primates. After millions of years, we finally migrated to the other continents of the world. The adverse climatic conditions during the Ice Age could have been the reason for the migration. It is notable that during this period, the population of our ancestors dropped from hundreds of thousand to about ten thousand, or near extinction.

Once the climatic conditions improved about 70,000 years ago, they started the migration upwards, around 50,000 years ago, reaching the areas of Europe (Neanderthals) and eastwards, up to Australia (Aborigines). During this period, a second migration was also noted to have occurred via inland, when they left the Tropics and heading towards the Middle East and Southern-Central Asia (Chinese). From here they also colonised northern Asia, Europe, and beyond.

Around 20,000 years ago, a small group of Asian hunters moved towards the East Asian Arctic. By about 16,000 years ago, migration came to North America via Asia when the ice melted and connected the pathway. From the north, it took about a thousand years for them to reach the tip of South America.

Important Milestones in the History of Our Planet That Refutes the Spontaneous Creation Theory

Time	Event
bya	*billion years ago*
mya	*million years ago*
ya	*years ago*
13.7 bya	The big bang occurs, and this universe is created.
13.6 bya	The first galaxies are formed.
12 bya	Our Milky Way Galaxy is formed.
4.57 bya	Our solar system is formed.
4.54 bya	Earth is formed by accretion.
4.45 bya	Moon is formed from pieces of earth.
4.28 bya	Water started condensing on earth.
4.0 bya	Life began—Prokaryotes.
3.5 bya	Photosynthesis started in smaller organisms.
2.3 bya	Oxygenated atmosphere developed.
2.0 bya	Aerobic respiration using free oxygen began.
1.5 bya	Eukaryotes appeared.
1.0 bya	Multicellular organisms are developed.
730–635 mya	There are two snowball earths.
542 mya	Cambrian explosion occurred. There is vast multiplication of hard-bodied life. The first abundant fossils are left behind.
450 mya	Bony fish are present in water.

380 mya	Insects and the first vertebrate land animals are found.
251 mya	Permian-Triassic extinction occurred, and 90 per cent of all land animals died due to cold and ice covering for a long period.
200 mya	Mammals appeared.
66 mya	Cretaceous- Palaeocene extinction occurred, and all dinosaurs perished due to meteorite impact in Mexico region.
7 mya	Hominines first appeared.
3.9 mya	First Australopithecus, the direct ancestor of modern Homo sapiens, is found.
3 mya	Homo erectus developed; the earth experienced another ice age.
700,000 ya	Fire was first used by hominids.
200,000 ya	The gfirst modern Homo sapiens and neanderthals appeared.
120,000 ya	Neanderthals appeared in Europe.
105,000 ya	People came in to forage for grass seeds, such as sorghum.
11,000 ya	Farmers and nomads began their activities.
6000 ya	Cities were built by tribes.
5000 ya	Kings and peasants came; writing was introduced.
250 ya	The Industrial Revolution started.
60–70 ya	Scientific discoveries were in full swing all over the world.

Chapter 5
The End of the Universe

The Significance of Dark Matter

Scientists have discovered that normal matter we know in the universe—the stars, the planets, rocks, oceans, gas clouds and the dust—when added together is only a small fraction, just 5 per cent, of the total mass of our galaxy. This matter is called "regular" or the "baryonic" matter as it was formed from the particles (baryons) in the quarks soon after the big bang.

The rest of the mass of the universe that cannot be seen is called the "dark matter" that was formed by something other than the baryons.

Dark Matter

The first evidence for the existence of dark matter emerged in 1930 through the work of the astronomer Fritz Zwicky. By 1970, other astronomers confirmed this when they discovered that the stars in the outer parts of the galaxies were orbiting their galactic centres faster than expected, and they apparently moved fast enough to escape their host galaxies. With this clue and other calculations as evidence, scientists concluded that dark matter appeared to surround galaxies in a halo, extending beyond the edges of the visible galaxies themselves, as well as existing in the space between galaxies in clusters.

The baryonic particles absorb radiation passing through them as they interact through gravity, nuclear forces, and electrostatic force. This interaction allows baryonic matter (stars) to emit light, and we can see them. But dark matter interacts only through gravity. Thus, it cannot emit light and is unable to be seen in detail other than as dark areas in space.

By measuring the ratio of hydrogen, which is the most common element in universe, and its isotope deuterium, scientists were able to calculate the baryonic matter in the universe. Deuterium only forms at very high temperatures in the early universe. The exact amount of baryonic matter created influenced the hydrogen/ deuterium ratio. From this ratio, the scientists found that the mass-energy of ordinary matter is only 5 per cent. The rest is dark matter, about 25 per cent, and dark energy, about 70 per cent.

However, dark matter is not antimatter. Scientists could not see the unique gamma rays in the dark matter that are produced when antimatter annihilated (crush) matter. The dark matter is found to be made up of other more exotic particles called weakly interacting massive particles (WIMPS).

The Significance of Dark Energy

There are three possible explanations for dark energy.

1. Albert Einstein described the constant energy filling the space homogenously as a cosmological constant, a vacuum energy, or the mass of empty space.
2. 2. Dark energy causes the drive force that accelerates expansion of the universe via a kind of generated repulsive force in the universe.
3. Dark energy may be the fifth force (quintessence) generated soon after the explosion that fills up the universe like a fluid.

However, dark energy is not yet a fully known form of energy because it is very dense and thought to permeate space. Dark energy is also not known to interact through any of the fundamental forces other than gravity. Since it is so dense, it cannot be detected in the laboratory as well. And unlike ordinary energy, dark energy does not get diluted with the expansion of the universe. This energy is supposed to accelerate the expansion of the universe in the early stages, and at the same time, it creates a repulsive gravity during the end of the universe, when all the normal energy is burnt out. More discoveries are yet to come about dark energy, but it plays its part during the end of the universe to complete the reverse process (reaction) back to big bang (creation).

However, it appears that dark energy is the other side of the coin to ordinary visible energy that we experience in the universe. Dark energy was created at the same time as normal energy was created, soon after the big bang. This is in accordance to the natural (plus and minus principle) phenomenon. Just a practical analogy being, "When you dig a pit on a flat ground, a small hill opposite the pit is formed next to the pit from the dug out material." To reverse this to the original state (equilibrium), the same material has to go back into the pit.

The Significance of Black Holes

A black hole is a region in space composed of densely packed matter with a gravitational pull so strong that nothing, not even the fast light (186,000 miles per second) can escape.

Black holes are supposed to be the cosmic vacuum cleaners or recycling machines of the universe. A black hole is a point in space where severe force of gravity pulls and sucks anything closer to it.

Following the big bang, visible and invisible forms of matter and energy were created in the universe. Early in the universe, the temperature was extremely hot and unevenly dense. The areas that were denser collapsed and created microscopic black holes due to higher gravity.

Larger black holes were formed in the centre of each galaxy during the formation of galaxy clusters. Later, when the stars were burnt off, more and more smaller stellar black holes were formed.

The black holes are not easily visible. But using special tools and technologies, they are detectable from their activities. These tools can show how the stars that are very close to a black hole act differently than those found further away.

The masses of objects in space are determined by the gravity they exert and the effects they produce on the matter around them. Using this principle and the equations (rotation curve) that describe the orbits of the stars nearby, the actual estimate of the mass of the black holes can be determined. Black holes can be as small as an atom, but the mass could be that of a mountain.

The largest black holes are about million times the size of our sun and are found in the centre of the galaxies. Our galaxy's super-massive black hole is called Sagittarius A.

The medium-sized black hole is called a "stellar," and its mass could be about twenty times that of our sun. These are formed when the core of a massive star burns out and collapses to form the supernova, or from an exploding star. Dozens of these are found within the Milky Way Galaxy.

Fortunately, at the moment, there are no black holes detected near our solar system. Our sun cannot become a black hole in the near future, and it is not big enough to form a powerful one.

But a black hole can grow bigger by sucking in additional matter, gases, and interstellar dust from its close surroundings and from cosmic background radiation. Two black holes, if they get too close, can merge to form one larger black hole.

A BLACK HOLE - Graphic

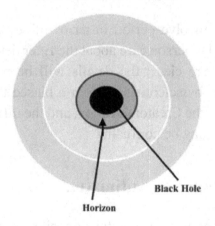

Black holes have no electric charge or angular momentum. The extreme gravity of a black hole can distort space, time, and light. Time runs slowly near the horizon and stops on the horizon of a black hole. There are about nineteen detected black holes in the Milky Way. The Sagittarius A, being the super massive one, with a mass of about 4 million times our Sun. The stellar black hole Cygnus is about fifteen times our sun's mass and is about 6,100 light years away from us. The others, like Aquila, are all smaller, between three and fifteen times of the mass of our sun.

It is important to note that the dominant activities of black holes will take several billions of years. Eventually, it is expected that only one big black hole will be left. The increased gravity and the density inside the last black hole will cause further shrinking to form the tiny dense singularity or the primeval body for the next cycle of a new universe.

Considering the concept of infinity to the universe, this is the best theory one can formulate from present scientific and historical facts.

Since all these involve period in terms of several billions of years, should a layperson care about the finer details too early? But the basics are clear; the details will be researched and discovered later by scientists. But for a biased theologian, the fundamentals of the Creator, creation, and the infinite condition of the universe may be confusing.

The Significance of Antimatter

Whenever matter was formed, the opposite—the antimatter— was formed with it (the two sides of a coin principle) under big bang conditions. Thereafter, antimatter stays separate without interacting with matter until towards the end of the cycle of the universe. The reason could be the energy and gravitational forces associated with the normal visible particles of matter. When conditions change, matter and antimatter have to bond again to the initial balanced state, like in a reverse chemical reaction. This occurs during the end of the universe (or big crunch) when all the energy in the universe gets exhausted or becomes unusable.

The basic difference between matter and antimatter is that they have opposite charges. In matter, we have the positively charged proton and negatively charged electron. But in antimatter, we have positively charged antiprotons and negatively charged positrons.

Matter and antimatter have subtle differences in helicity (spin) property. When matter travels in space, they make a left-handed spin, but antimatter has right-handed spin. This results in differences in the decay time (short), so we do not see lots of antimatter in the universe except when produced in the laboratory.

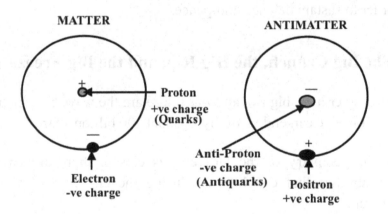

As such, antimatter is not expected to play a big role until the end of this universe. Hopefully, based on the instability, the antimatter is expected to collapse under gravity. However, it is expected to play some of its role (reverse action) in the last stages of our universe towards forming the singularity. This may take about 100 billion years.

One of nature's laws states that "To gain something you have to lose some other." It is a natural phenomenon when a neutral state is disturbed (split); it ends up with opposites that were bonded together. The matter and antimatter concept is also similar. Fission occurred at the beginning and bonding at the end, when the conditions changed. This is a natural phenomenon for equilibrium and one of the most intelligent mechanisms of nature. When conditions change, the reversal happens, and makes the system self-sufficient for materials.

Understanding this principle is very important for our lives too, especially as we strive to achieve balance after instability or problems. We must change the conditions appropriately in order to be back in peace. In fact, this is one of the simplest and most intelligent mechanisms we have to learn from nature in order to sustain balance and peace.

The Big Crunch, the Big Rip, and the Big Freeze

The big crunch, big rip, and big freeze are the ways by which our universe can end virtually in about 100 billion years.

If the geometry of the Universe is closed, there are two possibilities that can happen during the end part of this universe.

1. Burning stars are the main building blocks of our visible universe. When all the stars burn out and the whole energy of the visible universe is completely used or becomes unusable, there will be lack of repulsive energy, and gravity will also eventually seize. Then the expansion of the universe will have to slowly stop and collapse. At this time, all the matter

and debris in the universe will be sucked into black holes. In short, the reverse process to the big bang will start to end the cycle of this universe until the tiny singularity is formed for the next cycle.

2. If the universe is flat, the large amount of active dark energy left can make the universe continue the expansion slowly. But when the universe is collapsing, the protons and neutrons will disintegrate to positrons and electrons that, over time, will collide to annihilate each other.

3. The third possibility is the open universe, in which expansion can continue, meaning it will take much longer to end.

However, there must be an end to this universe when all the energy is completely used. The end will either be by fire or freeze, which may take up to 100 billion years. The finer details are to be researched and is a matter for scientists. Let us stay with the outline for the time being as our precious life is very short. Please refer the graph on page 00.

During the End of the Universe

The periods that we are working with in this regard is very, very long. Hence, it is hard to predict now the exact pathway to the end. However, as there was a beginning, there must be an end for our universe, until usable energy is again available in the universe. (Energy cannot be destroyed or created—Law of Conservation of Energy.)

For most people, the period of several billions of years may give a virtual eternity to the universe. When the events are in a cycle, there has to be a recycling in order to continue. Or new materials have to come from outside to restart the process, which is not an intelligent setup at this level.

The laws of conservation of matter and energy in thermodynamics complements this to say that matter and energy can neither be created nor destroyed and must go in cycles.

The events and products of this cycle easily answer the misunderstood question of nothingness from which the universe emerged. As a matter of fact, there cannot be any nothingness in a continuing cycle or a chain. There always has to be something small or big, visible or invisible in a cycle of events.

As mentioned earlier, after the big bang, tremendous energy came out with matter formed by fission and fusion reactions. With time, the space had to expand to accommodate various particles, objects, and activities of the universe. This expansion and the activities are progressing. One day they will hit the peak and naturally start to drop like a parabola when all the usable energy is used. So finally, this universe has to collapse and return to the initial state. This is called the big crunch, which ends in a tiny original body, the singularity that is to explode again and continue.

So the end is certain, but the finer details of the pathway are yet to be confirmed. As it is, there are two other possibilities, such as a big rip or a big freeze. Or all three of these can happen. Scientists are working on it, and it may be too early to describe and confirm.

Looking at the events in progress, the black holes, dark matter, and dark energy present will have to play important parts for the reverse process. Initially, fission occurred from singularity, and visible matter particles and energy came out. So at the end, the reverse action, fusion should happen. The used-up debris of visible matter and the unusable energy will get sucked up by the black holes and neutralised or annihilated with the dark matter and dark energy and then finally squeezed to form the tiny initial singularity for the next bang.

Black holes have already started functioning by sucking the debris of the cosmos using gravitational force. Inside the black holes, the debris is compressed and collapsed.

As mentioned earlier, eventually, two neighbouring black holes will merge into one, and a stronger and larger black hole will remain. This last black hole will end the universe, causing it to crunch, collapse, and compress further due to the gravity from the dark energy, to reverse the actions of the initial expansion and to form the tiny singularity for the next cycle of the universe.

Anyway, it is certain that this cycle will be completed and a new cycle for the next universe will begin. This will take several billions of years, and we need not worry too much now. A start means there has to be an end. It is similar to birth and death.

End of Universe in Open, Flat, and Closed Types

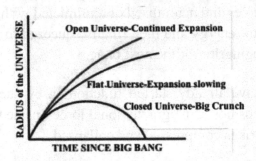

GRAPH TO SHOW POSSIBLE WAYS TO END

RADIUS of the UNIVERSE

Open Universe-Continued Expansion

Flat Universe-Expansion slowing

Closed Universe-Big Crunch

TIME SINCE BIG BANG

Other Types of Possible Universes and Their Ends

Anything that has a start should have an end (two sides of a coin). Having the big bang as the start of creation, the end has to be a reversal process of events where recycling is also involved. Since energy is being continuously used for most events, there must be a point when all the energy in the universe is used. This means the universe is not infinite but has a start and end. Thus, it is a closed type of universe.

There is a different view possible, that our universe is an open type, where our universe is supposed to continue the present course of expansion.

The third view could be that our universe is a flat type and will continue slowly forever.

In contrast, few have hypothesised a multiuniverse (multiverse) theory, where there is a possibility to have many universes like ours existing together. They are also called parallel universes, and physicists are divided as to their existence.

However, in our cases as common human beings trying to have a peaceful, happy life for a relatively short time (an average of eighty years), we need not get into the finer details and random hypothesis. Rather, we stay with what we can learn now and what we can do practically.

Chapter 6
Nature as a Whole Is the True God

Nature's Laws

In this world, there are many human-made laws (legal, religious, communal, cultural, and spiritual) made by various authorities. These laws are made and amended periodically to enforce law and order for discipline and social harmony in our society.

Most of these laws are based on common sense and derived from the unwritten laws of nature. Even scientific laws that control all the activities of the universe and others that are discovered by humans are originally nature's laws. For example, Newton's Laws of Motion was already functioning as a nature's law, but when discovered by a scientist, people considered it a scientific law.

Nature's laws are simple and logical. If understood and followed, they will guide all of us in living an easy and efficient life. When these laws are ignored, we directly or indirectly get into problems.

Nature's laws are automatic and set in such a way that there is no need for someone—God—to police, monitor, or keep records. Nature is simple, efficient, and works with the highest intelligence.

The several unwritten laws of nature, including those that govern the order and discipline of the universe, cannot be ignored. They require no monitoring; there are no loopholes. Nature's laws are fair and operate without favourites.

The following are some of nature's laws. -

1. Law of cause and effect: - There is a cause for every effect in the universe; it works similarly to the domino effect and continues as a chain of action.

2. Law of balance or equilibrium: The universe always tries to attain balance or equilibrium. To gain something you must lose another. For example, to gain wealth, you have to lose some of your time, energy, and so on.

3. Law of evolution: You are supposed to adapt and evolve according to the changes in the environment and competitions, or you may perish.

4. Law of desire: You get what you deserve, not what you desire. You reap what you sow.

5. Law of possession: The moment you possess something, you are possessed by it. For example, when you buy an expensive car, you are possessed by it, always worrying it may get damaged or stolen.

6. Law of attachment: Attached you lose, detached you gain. Attachment results in anger and sorrow when you lose what you are attached to.

7. Law of time: Time never stops, so live in the present before you lose that moment. The past is gone, and the future depends on the present.

8. Law of spiritual development: Spiritual development is inversely proportional to your worldly requirements.

9. Law of thoughts: Your thoughts become your actions, your actions become your habits, your habits become your character, and your character forms your destiny (life). This is an important law that controls our sins and successes.

10. The law of connection: Everything in the universe is connected. So life, too, is connected with nature, and the cause-and-effect principle is applicable all over.

11. Law of heredity: You have no choice at birth but to carry your parents' and grandparents' genes.

12. Law of birth: All those born must die one day.

13. Law of personality: What you think so you become.

14. Law of karma: You reap what you sow.

15. Law of knowledge: Knowledge proceeds from the known to the unknown.

16. Law of service to humanity: When you consume the resources of nature, you have to contribute back to nature in some form.

17. Law of success: No one can succeed without performing the appropriate action.

18. Law of progress and efficiency: Channel your energy with effort on the work to be more efficient.

19. Law of happiness and peace: You cannot attain peace and happiness unless your environment is happy and peaceful.

20. Law of wealth and prosperity: When you work with a spirit of renunciation, the world will reward you.

Why Nature Is the True God

Recent scientific and historical discoveries described in the previous chapters provide significant evidence to refute the spontaneous creation theory, which is centred on God as the Creator. To be frank, it is the whole universe, or nature, including its energy, that is the true and visible God or Creator. It is also clear that everything in this universe is dependent on some other cause and effect. Nothing can exist independently. The following brief supports this.

1. A visible nature, on its own, creates, sustains, and destroys, performing all the duties of an invisible God. Hence, there is no place or a need for a double. It has the required matter and the energy to operate automatically on a cause-and-effect principle. The events are in a cycle with no marked beginning or end. All the events and changes that are happening in the universe and on earth have reasons and follow certain laws. If God has created the universe, then we get a similar question: Who created God?

2. Nature has no assumptions or secret revelations like in various religions, where a single messenger from an unseen God or representatives of God (angels) descended from heaven to help selected groups of people. In contrast, nature

(the universe) is straightforward, observable, and researched by thousands of intelligent scientists for several decades.

3. The automatic cyclic events and recycling of materials in the universe demands that there is no need for another outside operator or a supervisor like God. That God monitors and judges us by our deeds is the religious concept of all religions. This reflects that God has only one job to do on a tiny planet with human species only. Further, what was God doing in the prehuman era? In the case of nature, the efficient laws of the nature govern the whole universe, including our planet directly and indirectly.

4. Nature is fair to all and does not discriminate between devotees and non-devotees or believers and non-believers. Nature cannot be bribed with praises, prayers, sacrifices, and offerings, like what we believe in religions and try to deceive the human-made gods.

5. Believing that our God are keeping account of our activities and monitoring us from heaven is another demerit to the Almighty. Practically, it is not possible in the huge universe when our planet is just a speck. But nature is all over, controlling the day-to-day activities of the universe. This includes all the lives on our planet in a fair way, not giving more importance to any group or individuals.

6. Being visible and known, nature eliminates the need for assumptions and myths. In addition, it does not try to explain unknowns via another unknown. For example, the religious gods are invisible and unknown, but people use other unknowns to explain God and his existence. Concepts

like heaven, hell, soul, fate, and rebirth are all unknown. No one has truly seen or experienced them personally.

7. Earth was formed about 4.5 billion years ago with a single elevated land mass and water surrounding it. But at present, we have seven continents with oceans surrounding each because of nature's effects, such as volcanic activities, earthquakes, tsunamis, cyclones, and storms. This is a big change that occurred gradually over about 3 billion years. The length of time required for the earth to evolve and change to sustain life refutes spontaneous creation theory.

8. As mentioned before, nobody can deny the fact that truth is just one and cannot be many. Unfortunately, we have too many faiths and divisions within the faiths, including many Gods to each of them. This is terribly confusing. In contrast, the whole nature is just the only one common and legitimate to all living things and performing all the activities we expect from a God of any religion.

9. If it was a loving God who created us, why would he create us with too many discriminative differences in colour, religion, culture, and other areas that make us suffer? Religions have no reasonable explanations for the destruction and suffering caused by uncontrollable natural disasters and other problems caused by fellow humans who create violence, wars, and genocides. They have no proper answers for births of children with disabilities and so on. In ancient beliefs, death and destruction, including natural disasters, were attributed to the wrath of the Gods, which does not fit into the honey-coated definitions given to their God.

Reasonable explanations for these disasters are possible only from the concepts of nature. Common sense would say, "You should be at the right place at the right time," to be trouble-free. This means right actions should be performed to get good results.

10. Who is responsible for innocent children born with permanent deformities such as Down's syndrome, autism, and Asperger's syndrome, limbless births, and so on, allowing them to suffer their entire lives?

Again, religions cannot provide reasonable explanations for these anomalies in creation other than to believe it as a punishment to the past sins.

Why would a loving, forgiving, and merciful Creator God allow this to happen to those who are yet to set foot on this earth and make them suffer their whole lifetime? Some beliefs rely on explanations that are unknown and try to prove with another unknowns, such as karma and fate. But nature explains this very well using genetics and other stress factors that affect the natural development of the foetus. Nature never displays miracles, unlike the religious beliefs, where God does miracles for selected individuals. If so, why are miracles not done for these innocent buds?

11. The five global extinctions in the history of earth, which eradicated about 90 to 96 per cent of living organisms and species every time, occurred as a result of natural disasters, such as snow covers and dust covers caused by the impact of meteorites and asteroids. It raises the question as to why God, as the Creator, would allow such destruction and restart creation five times.

12. All human-made laws (science, legal, social, cultural, and religious) are directly or indirectly derived from nature's laws. In fact, all scientific laws and others were already in force in the universe. The scientists discovered what already existed. For example, an apple when ripe always fell to the ground, even before Newton discovered the law or articulated the law of gravity in 1729. Thus, the laws of the nature are another affirmation for nature to be considered the true God.

13. As outlined previously, the reason earth is so finely tuned for living organisms to survive is because of natural features, like size, location from the hot sun, speed of rotation, tilt, and so on. The environment naturally changed from very hot to optimum over a billion years. If God were the Creator, he would have finely tuned our planet with a suitable environment from the outset, as per his spontaneous creation policy. This also raises the question of why the other solar planets were not used for life. And why any other was not suitably tuned for creation by God.

14. The very long time, billions of years, taken for evolution of both non-living and living things is also important evidence that refutes sudden spontaneous creation theories expounded by the religions. Nature, however, acts gradually and slowly, and no magic is involved.

The big bang theory and the natural selection theory show that all the major events and changes in the universe are automatic. The events occur through a cause-and-effect mechanism, like in a domino setting in a chain. A cause is followed by the effect, and this effect becomes the cause for the next effect, and so on. The time involved in this process is so long that it is not easy for our brains to comprehend, let alone understand the sequence without a supercomputer.

Summary

1. Nature creates, sustains, destroys, and recycles.
2. Nature has no beginning or end because it is cyclic.
3. Nature is omnipotent, omnipresent, and omniscient.
4. Nature is independent, fair, and self-sufficient.
5. Nature has efficient natural laws that require no operator or monitor.
6. Nature does not answer questions via unknowns.
7. Prayers, praises, and bribes do not sway nature.
8. Nature needs no offerings or sacrifices.
9. Nature is one and common to all, can unite, and is the single and true.
10. Nature answers any questions on life with genuine reasons, and understanding it makes life easy.

Chapter 7
Our Lives

Why Are We Born and What Is the Real Purpose of Life?

Our births look intentional but are basically accidental in so many ways. They are accidental in that the creators of life, the parents, have no ability to shape the outcome. In fact, it is really the combination of our genes that play the role of the Creator.

Despite the fact that our births are surrounded by elements of chance, randomness, and lack of control, we often seek meaning and purpose to our births and lives.

To put it simply, we are born to live as long as possible without suffering. Luckily for most of us, survival is not a problem, not like what our ancestors went through. Survival is better organised now than before, so we are at the next step, seeking comfort and peace.

However, most of us fail in life and never realise that this opportunity to live is fairly short. We also must understand early that in the scheme of the universe, we are just a speck and can live for only a fleeting moment. There is no time to waste.

The foundation for this truth begins with the realisation that nature is the Creator, or the true God. There are several proofs available to support the big bang theory and natural selection. If nature is God, it opens up a tremendous possibility to take responsibility for our lives, neither blaming nor praising God.

Instead, by understanding the laws of nature, we can determine the most appropriate action at any given time to avoid or solve problems.

Letting go of the needing to have a human-made God to impress, beg, pray, praise, blame, and revere we save a lot of time and energy. This allows the species on the highest ladder of evolution to live better with human values.

Actions like blaming, praising, pleasing, or trying to impress God not only downgrade the Almighty but also reveal our ignorance. Many of us spend so much of our lifetimes praying and visiting places of worship without performing appropriate actions or taking responsibility. We see successful people belittle their successes to divine grace or blessings. We fail to realise that real long-term success in life comes from having the confidence and wisdom to take the most appropriate action for the situation.

The big bang theory and the evolution theory, where nature is the Creator, can help us answer some existential questions of life: Why are we born? What is the purpose of life? What should I do to have a peaceful and happy life?

The Real Purpose of Life

People define their life's purpose based on their level of understanding and their priorities. Many believe that life is mainly to earn money and wealth. These people spend most of their lifetimes away from home, working long and hard, spending little time with families and friends. Remember that no one has enough in this world, including a millionaire.

Others focus on academic education and give their intellects highest priority. There are those looking for power and position. There are many who believe that their life's purpose is to provide for their families. There are also those who like to spend their lives hoping to leave their names behind in history. Some spend their lives trying to create personal records and achievements. Religious individuals focus on finding heaven or liberation.

Basically, life is survival. But fortunately, we have evolved higher, are civilised enough, and have no problem when it comes to surviving. So we now seek comforts and superiority. The truth is that life is time, and it cannot be stopped or paused. We are given a single chance to live, and it is brief too. We can't afford to have trial-and-error checks as well. Hence, some planning is required to choose the right priorities and keep moving on. It is essential that you live in the present. Living a complete and balanced life is very important, and so is living a life without regrets. Regretting later that you have failed to live better or asking for a second chance means that you have wasted the life you have. If we are conscious, cool, and steady, our actions will be productive.

Below are a few chosen styles of life from different perspectives: If,

- Life is a challenge, meet it.

- Life is a gift, accept it.

- Life is an adventure, dare it.

- Life is a sorrow, overcome it.

- Life is a tragedy, face it.

- Life is a duty, perform it.

- Life is a game, play it.

- Life is a mystery, unfold it.

- Life is a song, sing it.

- Life is an opportunity, take it.

- Life is a journey, complete it.

- Life is a promise, fulfil it.

- Life is a love, enjoy it.

- Life is a beauty, praise it.

- Life is a spirit, realise it.

- Life is a struggle, fight it.

- Life is a puzzle, solve it.

- Life is a goal, achieve it.

If all the above are mixed in the right proportions and we live our lives without hurting others, our lives become successful. The quality of our lives depends on how we are peaceful and joyful. We are supposed to live in reality rather than yield to all our thoughts and emotions. Most of us fail in life when we compare our lives with others. A very thin line exists between need and greed for life, and care should be taken not to cross it. Further, in life we should be able to differentiate the things that we can do and those that cannot be done. Not doing what we could do and trying to do what cannot be done are disastrous.

Conclusion

No one can deny that in the last sixty years or so, the marvels of modern science have changed our lives quickly and dramatically, and almost in an unimaginable way.

Modern science has also provided us with the updated knowledge and evidence about creation and the true Creator. This true knowledge and the truth have not yet reached many as it is a new finding. But it is available to one and all if we are keen to review and research our thoughts, philosophies, and beliefs without prejudice.

Yet so many of us tightly hold on to the ancient beliefs about God and creation that were expounded at a time when there was neither modern science nor the technology and the education to know better.

Evolution is all about adaptability and flexibility in learning and enhancing our perceptions. This allows species to genuinely survive, thrive, and improve.

When someone understands and believes in nature as the true controller of the universe, the individual will not be troubled by problems of a mundane life. Human are not at all special in this vast universe. We are supposed to adjust according to nature and the environment.

On the other hand, when someone believes God is sitting in heaven and created us, God is being used as an insurance policy, to seek favours and shortcuts to solving problems are expected. This wrong belief creates problems and keeps one asking silly questions such as, "Why me, God?"

People who thoroughly understand about the true creation and the Creator will have no fear, ego, anger, enmity, sorrow, worries, depression, or suicidal thoughts as they understand natural events.

We have to remember that this life is very short. It is a rare privilege given to us to be in the highest position in the evolution. Our ancestors struggled for millions of years with the environment and predators to survive to reach this peak level in our species. And then they passed it on to us. So this life has to be lived thoroughly, rather than getting fooled by any unrealistic beliefs and false hopes.

All who believe in God are not blessed. Just look at the majority of innocent people who blindly visit places of worships regularly, year after year, coming out with no blessings. It is a pity they continue doing the same again, only to end up with nothing but grief and agony and have to be helped.

The ages of the universe and our planet are other important factors to understand the history. According to many religions, the age of our planet never exceeded 15,000 to 20,000 years. Unfortunately, in the ancient time, our ancestors did not have advanced technology and tools to know most of the facts. Therefore, they had to assume things and would have chosen a spontaneous creation theory and a super-God to do their tremendous work.

We have been assuming and seeking an unknown and unseen via another unknown when the true controller the nature is just in front of us!

Blind faith is similar to a beautiful dream. It may provide solace for a short time period. But as life gets more complicated, this will lead people in a wrong direction with confusion and helplessness.

Obviously, when someone is successful, he or she has taken some appropriate actions. This is an important nature's law. Wrong faith often leads individuals to relinquish their responsibilities and to rely on praising or blaming God.

I hope the various facts and evidence given in this book will support the theory of natural creation. Nature is the ultimate non-transparent almighty, that freely flowing self-sufficient controller of the universe. For people to understand and change beliefs, it may take two to three generations. It is very difficult to change cultures and traditions immediately, but at least we should cull the obvious false beliefs, time-consuming senseless practices, and extravagant spending on worship places. It is important to remember that we are obliged to direct our future generation on the right path and unity.

Other Publications

Poultry Production and Common Diseases for Farmers and Students. (Hand Book for Reference).

About the Author

Pon Satchi is formerly a research assistant at the University of Sri Lanka, a government veterinary surgeon, served in the Ministry of Agriculture in Sri Lanka. He also served as a poultry production consultant for Omeri Farms and Feed Mills in North Yemen.

Dr Pon Satchi comes from a devoted Hindu family and has a liking to follow spiritual life. He attended various spiritual group meetings and lectures, including Sai, Chimaya, Vedanta, Sadguru, and Bible study without discrimination in order to seek the truth for a happy and peaceful life. After retirement, he went into self-study and research, reading books and scientific internet publications with an open mind. Now he wants to share his recent discoveries about the truth behind creation and the Creator with other freethinkers. He is not an atheist or a canvasser.

Printed in the United States
By Bookmasters